Development of New Optical Techniques and Methods for Basic Biology and Biomedical Applications

Development of New Optical Techniques and Methods for Basic Biology and Biomedical Applications

Guest Editor

Anna Cleta Croce

MDPI

Basel • Beijing • Wuhan • Barcelona • Belgrade • Novi Sad • Cluj • Manchester

Guest Editor
Anna Cleta Croce
Institute of Molecular
Genetics
Italian National Research
Council (IGM-CNR)
Pavia
Italy

Editorial Office
MDPI AG
Grosspeteranlage 5
4052 Basel, Switzerland

This is a reprint of the Special Issue, published open access by the journal *Biology* (ISSN 2079-7737), freely accessible at: https://www.mdpi.com/journal/biology/special_issues/Optical_Techniques_Methods.

For citation purposes, cite each article independently as indicated on the article page online and as indicated below:

Lastname, A.A.; Lastname, B.B. Article Title. *Journal Name* **Year**, *Volume Number*, Page Range.

ISBN 978-3-7258-5035-8 (Hbk)
ISBN 978-3-7258-5036-5 (PDF)
https://doi.org/10.3390/books978-3-7258-5036-5

Cover image courtesy of Anna Cleta Croce

Contents

About the Editor

Anna Cleta Croce

Anna Cleta Croce, Istituto di Genetica Molecolare, IGM-CNR, Pavia, Italy. Interests: fluorescence, induced or natural (autofluorescence), spectroscopy (UV-visible; conventional or via fiber optic probes), microspectrofluorometry, and imaging. Anna Cleta Croce is mainly devoted to the characterization of natural biomolecules acting as endogenous fluorophores to develop autofluorescence based procedures for their in situ, real time detection and estimation, similarly to an in situ biochemical analysis. Her research activities have been supported by many projects, funded by the CNR, Italian Ministry of University and Research, "Cariplo" foundation and Italian Lombardy Region, serving also in coordination, administration and management. According to the term "optical biopsy", autofluorescence-based procedures have been applied to assess the morpho-functional conditions of normal or diseased organs and tissues. Great interest has been given to experimental hepatology, for the real-time, in situ monitoring of liver functionality under conditions such as hypoxia/reoxygenation in surgery or during preservation for transplantation, or for the detection of liver metabolic disorders. Recently, Anna has also been interested in entomology—assessing the optical properties of mosquitoes as basic research, and developing optical methods to support control strategies of vectors of pathogenic viruses. Her service as Guest Editor of the Special Issue was inspired by her unceasing goal of spreading awareness of the extremely ample potential of biophysical methods in biology and biomedicine, from basic studies to analytical, diagnostic, and therapeutic applications. She is the author/coauthor of more than 120 articles in peer reviewed Journals and book chapters, and about 300 contributions to national and international scientific meetings.

biology MDPI

Article

Enhanced TfR1 Recognition of Myocardial Injury after Acute Myocardial Infarction with Cardiac Fibrosis via Pre-Degrading Excess Fibrotic Collagen

Wenwen Yang [1,2,†], Yueqi Wang [2,†], Hongzheng Li [1], Feifei Liao [1], Yuxuan Peng [1], Aimei Lu [1], Ling Tan [1], Hua Qu [1], Linzi Long [1,*] and Changgeng Fu [1,*]

1 Graduate School, China Academy of Chinese Medical Sciences, Beijing 100091, China
2 CAS Key Laboratory of Molecular Imaging, Beijing Key Laboratory of Molecular Imaging, The State Key Laboratory of Management and Control for Complex Systems, Institute of Automation, Chinese Academy of Sciences, Beijing 100190, China
* Correspondence: qixiang830803@163.com (L.L.); fucgbs@163.com (C.F.); Tel.: +86-010-62835035 (C.F.)
† These authors contributed equally to this work.

Simple Summary: After myocardial infarction (MI), myocardial tissue undergoes a series of complex biological reactions, among which fibrosis is particularly critical. This process leads to troublesome collagen deposition, which not only affects the normal function of the myocardium but also poses a serious challenge to conventional myocardial injury imaging techniques. Especially in the metabolism of contrast media, the uneven distribution caused by fibrosis greatly reduces the accuracy of imaging. To address this problem, we propose an imaging method for myocardial fibrosis processing after myocardial infarction, aiming to improve the accuracy of myocardial injury imaging. Firstly, a combination of collagenase I and human serum albumin (HSA-C) was used to perform deep clearance of myocardial fibrotic collagen. Collagenase I is an enzyme that specifically acts on collagen fibers and can effectively degrade collagen fibers in the myocardium. However, human serum albumin serves as a carrier to help collagenase better distribute in myocardial tissue. After the successful clearance of fibrotic collagen from the myocardium, optical contrast agents targeting the transferrin receptor were used to precisely localize the damaged myocardium. The expression of the transferrin receptor significantly increased after myocardial injury, so this contrast agent can specifically identify the injured myocardium. This precise localization capability provides an important basis for subsequent imaging techniques. With the use of near-infrared fluorescence imaging, we can precisely track the location of myocardial injury. In the area of myocardial damage, the fluorescence signal is significantly enhanced due to the specific binding of the contrast agent, to achieve accurate tracking of the location of myocardial injury. This pretreatment strategy not only effectively eliminates the background interference caused by contrast agent deposition in fibrotic tissue, but also improves the targeting efficiency of contrast agents in cardiac lesions.

Abstract: The fibrosis process after myocardial infarction (MI) results in a decline in cardiac function due to fibrotic collagen deposition and contrast agents' metabolic disorders, posing a significant challenge to conventional imaging strategies in making heart damage clear in the fibrosis microenvironment. To address this issue, we developed an imaging strategy. Specifically, we pretreated myocardial fibrotic collagen with collagenase I combined with human serum albumin (HSA-C) and subsequently visualized the site of cardiac injury by near-infrared (NIR) fluorescence imaging using an optical contrast agent (CI, CRT-indocyanine green) targeting transferrin receptor 1 peptides (CRT). The key point of this strategy is that pretreatment with HSA-C can reduce background signal interference in the fibrotic tissue while enhancing CI uptake at the heart lesion site, making the boundary between the injured heart tissue and the normal myocardium clearer. Our results showed that compared to that in the untargeted group, the normalized fluorescence intensity of cardiac damage detected by NIR in the targeted group increased 1.28-fold. The normalized fluorescence intensity increased 1.21-fold in the pretreatment group of the targeted groups. These data demonstrate the feasibility of applying pretreated fibrotic collagen and NIR contrast agents targeting TfR1 to identify ferroptosis at sites of cardiac injury, and its clinical value in the management of patients with MI needs further study.

Keywords: myocardial infarction; human serum albumin; collagenase I; transferrin receptor 1; near-infrared fluorescence imaging

1. Introduction

Acute myocardial infarction (MI) is a severe clinical emergency in which necrotic areas are replaced by fibrous tissue due to the limited regenerative capacity of cardiomyocytes, resulting in ventricular remodeling with severe effects on cardiac function [1]. Myocardial remodeling after MI is an important contributor to cardiac dysfunction, increased complications, and increased mortality, and myocardial fibrosis plays a key role in this process [2]. Therefore, early and effective prevention and treatment of myocardial fibrosis after MI are of great significance. Myocardial fibrosis refers to the excessive synthesis and deposition of extracellular matrix in the myocardium, which is mainly composed of type I and type III collagen. Oversynthesis of these collagens reduces myocardial compliance and consequently, cardiac function [3,4]. The occurrence of myocardial fibrosis after MI is not only related to the necrosis of cardiomyocytes, but also to the expression of a variety of cytokines and the activation of signaling pathways [4]. For example, hypoxia, inflammatory response, and oxidative stress may promote the occurrence of myocardial fibrosis. At present, the treatment methods for myocardial fibrosis after MI mainly include drug therapy, interventional therapy, and stem cell therapy. Drug therapy is commonly used, but the therapeutic effect is limited. Interventional therapy can improve myocardial perfusion, but it is not effective in the treatment of myocardial fibrosis [5]. Stem cell therapy is an emerging treatment method that promotes the regeneration and repair of cardiomyocytes by transplanting stem cells to improve cardiac function, which needs to be further verified [6]. At the same time, cardiac fibrosis may lead to changes in the density and blood flow distribution of the myocardial tissue. These changes may affect the distribution and metabolism of the imaging agent in the myocardium, thereby affecting the results of imaging. For example, in positron emission tomography/computed tomography (PET/CT), myocardial fibrosis may affect the uptake and clearance rates of the glucose metabolism of imaging agents in scar tissue, leading to inaccurate measurements of glucose metabolism rate [7]. Cardiac fibrosis may also affect the elasticity of myocardial tissue. Due to the different elastic coefficients, myocardial scar tissue and normal myocardial tissue will undergo different deformation when subjected to the same pressure or strain. This different deformation may lead to changes in the motion pattern of the myocardial tissue, which may affect the measurement results of techniques such as echocardiography [8].

Ferroptosis is not apoptosis or necrosis in the traditional sense, but a special mode of cell death. During this process, the normal balance of iron metabolism of cells is disrupted, leading to loss of cellular structure and function. It plays a crucial role in the development and progression of several cardiovascular diseases, such as MI, myocardial ischemia/reperfusion, and arrhythmia [9,10]. The cells facilitate the uptake of iron through transferrin (Tf), a protein that binds to iron, and its receptor 1 (TfR1). This process is initiated by the interaction between the two, leading to the endocytosis of complexes containing iron [11,12]. Changes in the expression of TfR1 and the disruption of iron transportation from the cytoplasm to mitochondria result in augmented mitochondrial iron levels, which are linked to ferroptosis subsequent to cardiac injury [13]. Therefore, TfR1 is expected to be a biomarker for the specific detection of MI-induced ferroptosis because it is associated with MI-induced damage. However, existing detection methods are insufficient to detect abnormal expression of TfR1 non-invasively and sensitively in cardiac injuries. New technical means are urgently needed to overcome this limitation. Molecular imaging can detect MI at the molecular and cellular levels and has become an important tool for diagnosing and treating MI. Advanced molecular imaging technology, including PET/CT [14] and magnetic resonance imaging (MRI) [15], has been designed to visualize changes in MI. Near-infrared (NIR) fluorescence molecular imaging has been utilized for visualizing cardiac injuries [16]. However, it is worth noting that the acquisition times for MRI and PET/CT were relatively long. Consequently, there is a demand to establish highly sensitive imaging techniques that can accurately detect ferroptosis in myocardial fibrosis. A TfR1-targeted optical contrast agent was then used to specifically "illuminate" the MI focus using NIR fluorescent molecular imaging, and experimental procedures have been detailed in Scheme 1. Through our investigation, we found that the deposition and degradation of fibrotic collagen closely affected background signal interference in fibrotic tissue and uptake of TfR1-targeted contrast agents specifically designed to target MI within injured tissues. In the pathological process of myocardial fibrosis after MI, we first used human serum albumin-carried collagenase I (HSA-C) to degrade fibrotic collagen and reduce contrast agent retention in the cardiac fibrosis. The combination between HSA-C and TfR1-targeted peptides (CRT) fluorescence probe distinguished the sites of ferroptosis in myocardial injury accurately, and its normalized fluorescence intensity was 1.21-fold compared to that of CRT-ICG (CI) alone.

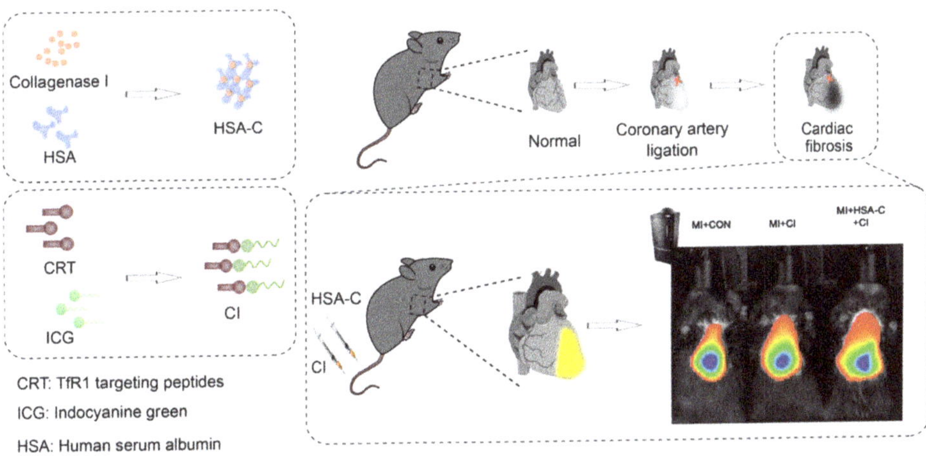

Scheme 1. Schematic of NIR fluorescent imaging for MI using HSA-C and the targeted contrast agent CI. In the process of myocardial fibrosis induced by myocardial infarction, synthetic HSA-C helps to promote the degradation of fibrotic collagen, which helps to reduce the non-specific retention of the contrast agent in the heart caused by cardiac fibrosis. Western blot assay showed that the expression of ferroptosis TfR1 was increased in the infarcted region (Figure 1A) and immunohistochemical staining showed that TfR1 was highly expressed at the site of myocardial infarction (Figure S2). Based on this, we designed and synthesized an optical biocontrast agent that combines the TfR1-targeting ligand CRT with the fluorescent dye ICG (CI) to specifically target TfR1 at the site of cardiac injury. During the experiment, we first used caudal vein injection of HSA-C to degrade the fibrotic collagen. HSA-C has a high specificity and affinity, and can accurately recognize and degrade fibrotic collagen, thereby alleviating the process of myocardial fibrosis. This step set the stage for subsequent experiments. Next, we employed an optical contrast agent targeting TfR1 at the site of cardiac injury. This optical contrast agent is highly specific and is able to bind tightly to the site of cardiac injury, allowing clear visualization of the site of injury. This step is to better understand the extent of ferroptosis in myocardial injury to provide a basis for subsequent therapy. After completing the above two steps, we used NIR fluorescence imaging to accurately display the MI lesion site. Near-infrared fluorescence imaging technology has high sensitivity and can clearly show ferroptosis in MI lesions. With this technique, we can assess the extent of MI lesions more accurately, and provide new ideas and methods for the diagnosis of myocardial infarction.

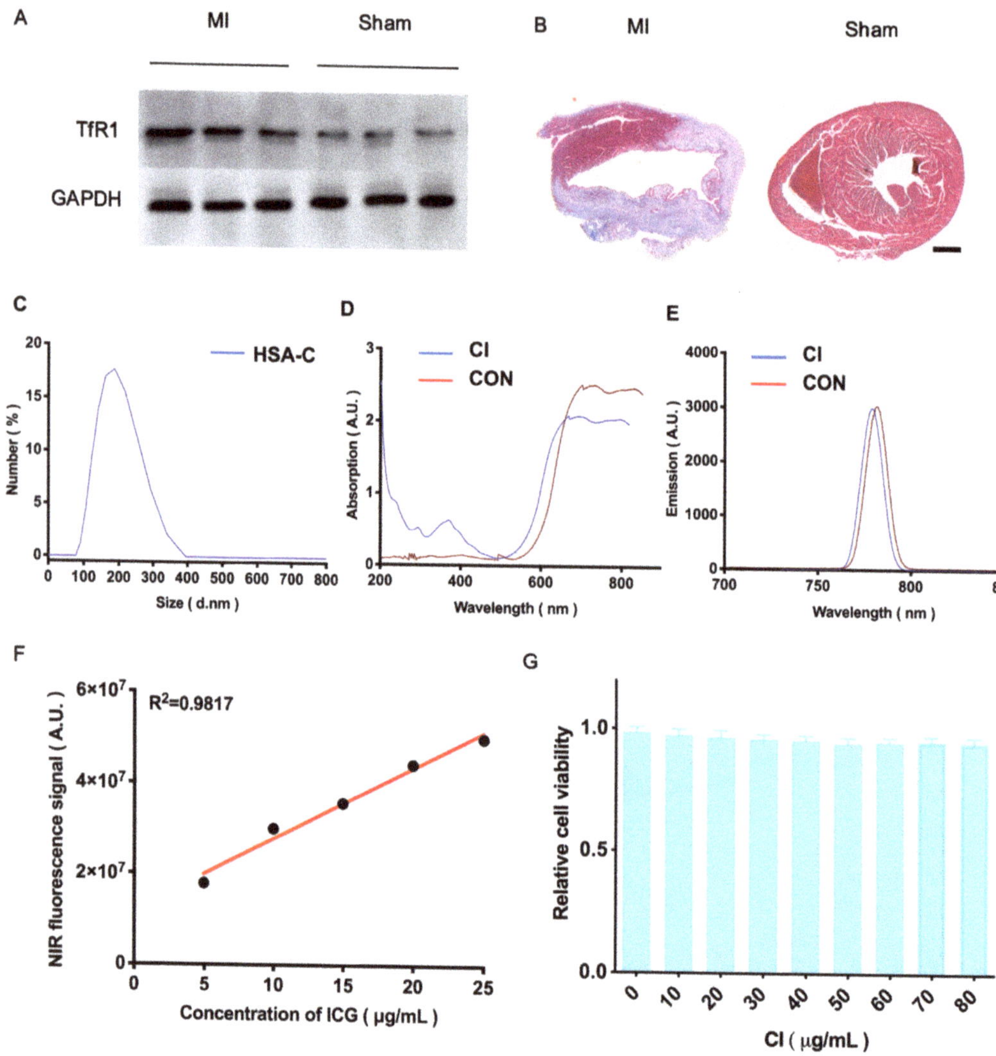

Figure 1. Characterization of the HSA-C and CI. (**A**) Expression of TfR1 in the MI mouse model and Sham group (MI vs. Sham) was detected via western blotting. (**B**) The Masson staining images of MI group model and Sham group (MI vs. Sham; Scale bar = 100 μm). (**C**) The hydrated particle size of HSA-C. Absorbant (**D**) and fluorescence (**E**) spectra of CI and CON. ICG-NHS was used as the CON. (**F**) The intensity of CI's fluorescence is enhanced as the concentration of the fluorophore increases when examining the IVIS spectrum. (**G**) After 24 h of treatment with different CI concentrations, the viability of H9c2 cells was assessed. (n = 6; one-way ANOVA).

Myocardial fibrosis can significantly affect the identification of myocardial injury, and degradation of excessive fibrotic collagen can make the identification of myocardial injury more intuitive, so HSA-C is the key to solving this problem. At the same time, the development of functionalized molecular contrast agents with high sensitivity and specificity is expected to accurately identify cardiac damage after MI.

2. Materials and Methods

2.1. Synthesis and Characterization of HSA-C

The HSA-PEG-Maleimide (HSA-PEG-Mal) polymer was self-assembled using a method previously described in the literature [17]. To summarize, HSA-PEG-Mal was dissolved in dimethyl sulfoxide and gradually added to deionized water. The resulting solution underwent dialysis against deionized water for 24 h and was subsequently centrifuged. The supernatant was collected and stored at 4 °C. To thiolate collagenase I, the Traut reaction was employed by introducing 5 × Traut to the collagenase I solution (5 mg/mL in phosphate-buffered saline [PBS]) and stirring the mixture at 25 °C for 1 h. The obtained product, namely thiocollagenase I (SH-Collagenase I), underwent purification using a desalting column (Thermo Fisher Scientific, Waltham, MA, USA). Next, HSA-PEG-Mal was reacted with SH-collagenase I at 25 °C overnight to prepare pre-C-HSA. Then, Mal-PEG-COOH was added and reacted at room temperature for 4 h. SH-PEG-COOH was added to the solution and reacted at 25 °C for 4 h. The HSA-C solution was obtained and centrifuged for 30 min. The supernatant was removed, and the pelleted polymer was resuspended in PBS and stored at 4 °C. Using a Zetasizer Nano (Malvern Instruments, Malvern, UK) to characterize the hydrated particle size distribution and zeta potential.

2.2. Synthesis of CI

The Tfr1-binding peptide CRTIGPSVC (CRT, GL Biochem Co., Ltd., Shanghai, China) [18] underwent the labeling process by conjugating with indocyanine green-N-hydroxysuccinimide ester (ICG-NHS; Solarbio Co., Ltd., Beijing, China) at room temperature. Specifically, the separate dissolution of ICG-NHS and CRT in dimethyl sulfoxide (Solarbio Co., Ltd., Beijing, China) took place. The resulting mixture underwent overnight agitation in the absence of light, followed by a 24-h dialysis period to eliminate any unbound small molecule peptides CRT. Ultimately, the lyophilization process was conducted to acquire CRT-ICG.

2.3. Characterization of HSA-C and CI

To conduct detection, CI was dissolved in deionized water. The Zetasizer Nano ZS (Malvern, UK) was employed at 25 °C to measure the hydrodynamic diameters of HSA-C. For optical absorption analysis of the ICG-NHS (ICG-NHS was used as the CON) and CI, a UV–Vis–NIR spectrophotometer from Shimadzu (Kyoto, Japan) was utilized. To measure CON and CI's emission and excitation spectra at 25 °C, a fluorescent spectrophotometer F-7000 from Hitachi (Tokyo, Japan) was employed. Fluorescence intensity analysis of CI at various concentrations was carried out using an IVIS spectrophotometer from PerkinElmer (Waltham, MA, USA).

2.4. Cell Counting Kit 8 (CCK8) Testing for Cytotoxicity

H9c2 cardiomyocytes were purchased from Wuhan Pronosay Life Technology Co., Ltd. (Wuhan, China), and all were species identification correct cells. H9c2 cells were cultured in a high-glucose medium (Dulbecco's modified eagle medium, DMEM) containing 10% fetal bovine serum (FBS) (Gibco, Emeryville, CA, USA) and 1% penicillin–streptomycin (Sevier, Wuhan, China) at 37 °C in a 5% CO_2 environment provided by the Heracell system from Thermo Fisher Scientific. Fluid exchanges were performed every 48 h. Cells within 10 passages were selected for experiments to ensure cell viability. When the cell density reached 90%, the cells were digested with 0.25% EDTA-containing trypsin, a complete medium was used to abort the cell digestion process, and the old culture medium was removed. The cells were resuspended in a complete culture medium, seeded in a 96-well plate at a density of 2×10^4 cells/well, and incubated at 37 °C for 24 h. To investigate the effects of CI, the probes were diluted in a special medium above. The concentration of the probe varied from 0 to 80 μg/mL, increasing by 10 μg/mL each time. Both the CI group and the control medium group were included in the experiment. Following the 24 h incubation, the cells were washed thrice using PBS. Then, 10 μL of CCK8 was added and the cells were incubated in the special medium for an additional 2 h. Measuring the optical density (OD) values at 450 nm by a multimode microplate reader from BioTek, Winooski, VT, USA. The results were analyzed as a percentage of the control pore value.

2.5. Ethics

The animal studies in this research were performed in compliance with the guidelines outlined by the Institutional Animal Care and Use Committee of Xiyuan Hospital, China Academy of Chinese Medical Sciences, and ethical review approval (2021XLC027) was obtained.

2.6. MI Animal Model Establishment

In this study, we utilized 7-week-old male C57BL/6N mice (Beijing Vital River Laboratory Animal Technology Co., Ltd., Beijing, China). The animals were kept in SPF-grade animal houses at 20–25 °C and 40–70% relative humidity. Mice were housed in cages with 5 mice per cage. The animal house was kept quiet, and mice were fed and watered ad libitum, with a 12/12 h cycle of light. The model of MI was established by anesthetizing 60 mice with 2% isoflurane. To expose the myocardium, a precise incision was made on the left chest of the mouse between its third and fourth ribs, followed by careful separation of the muscles. Roughly 2 mm below the junction of the left atrial appendage and conus arteriosus, the ligation of the anterior descending branch of the coronary artery took place, and no ligation for the Sham group. After the surgery, the heart was returned to the heart cavity and the skin was stitched up. The MI mice were divided into three groups randomly after the electrocardiogram confirmed ischemia: (1) MI + CON, (2) MI + CI, and (3) MI + HSA-C + CI.

2.7. Western Blot

The RIPA protein extraction reagent was precooled and the protease inhibitor cocktail was added. The heart affected by the disease was sectioned into smaller pieces. The tissue was homogenized by 10% at 30,000 rpm for 2 min and lysed by shaking at 4 °C for 30 min. Centrifugation was performed at 13,000 rpm (4 °C) for 10 min. The resulting liquid was collected by centrifugation to obtain the supernatant. The supernatant was used for protein quantification and protein sample preparation. The concentration of protein in the supernatant was determined using the Pierce BCA Protein Assay Kit. To begin the electrophoresis process, a G protein solution of 20 μL was mixed with a solution of SDS (sodium dodecyl sulfate) loading buffer. A polyacrylamide gel with a composition of 4% to 15% was used for the separation of the protein molecules. Electrophoresis proceeded at 100 V for a designated amount of time ranging between 90 to 120 min. After the gel no longer showed any traces of the bromophenol blue dye, the electrophoresis halted, and the protein subsequently moved onto the PVDF (polyvinylidene difluoride) membrane. Treatment of the PVDF involved placing it in the solution sealed at 25 °C for 1 h, and allowing it to remain overnight at a temperature of 4 °C. This solution contained a diluted primary antibody that was used for specific protein detection (dilution ratio of 1:1000). A decolorization shaker was utilized to rinse the PVDF membrane three times with tris-buffered saline with tween-20 (TBST), with each rinse lasting 15 min at a temperature of 25 °C. To proceed further, the PVDF membrane underwent treatment with a secondary antibody that was diluted at a ratio of 1:5000. This treatment took place for a span of 1 to 2 h at a temperature of 25 °C. Subsequently, the membrane was subjected to three additional washes using TBST on a decolorization shaker, maintaining a temperature of 25 °C. Lastly, a chemiluminescence reaction detection process ensued. The resulting data were then meticulously recorded and subsequently analyzed utilizing Image J software (1.53k).

2.8. Masson Staining

Mouse myocardial tissues were washed three times with PBS and subsequently fixed with 4% paraformaldehyde solid solution. After complete fixation, the tissue was embedded using the paraffin embedding technique and then cut into 5 μm thick sections using a paraffin tissue microtome; paraffin sections were deparaffinized to the distilled water stage according to routine procedures. Weigert iron hematoxylin staining solution can be prepared by mixing reagents A1 and A2 (Solarbio Science & Technology Co., Ltd., Beijing, China) in a 1:1 ratio. This stain was dropped onto the sections to cover the sections and stained for 5–10 min. After completion of staining, sections were washed using distilled water to remove excess staining solution. Next, acidic ethanol differentiation solution was added dropwise for 5–15 s, followed by a 30-s wash with distilled water. Staining was performed using Ponceau-fuchsin staining solution for 5–10 min. In the staining process, the weak acid working solution should be prepared according to the ratio of distilled water: weak acid solution = 2:1. A weak acid working liquid drop was added to the section and the wash time was 30 s. After the excess liquid was poured off, phosphomolybdic acid solution was dropped for 1–2 min. It was then washed again with a weak acid working solution for 30 s. After the excess liquid was poured off, aniline blue staining solution was dropped and stained for 1–2 min. This was followed by another 30-s wash using a weak acid working solution. Finally, a rapid dehydration was performed using 95% ethanol for 2–3 s. This was followed by dehydration using absolute ethanol twice for 5–10 s each. The transparencies were then treated twice with xylene for 1–2 min each. Finally, neutral gum was used for sealing. Microscopy, image acquisition, and analysis were subsequently performed.

2.9. NIR Fluorescence Imaging

Before fluorescence imaging, the hair on the chest of the mice needs to be shaved, because the black mice themselves produce fluorescence self-absorption. The mice were anesthetized with a 2% isoflurane–air gas mixture for the entire imaging procedure. Mice injected with optical contrast medium were imaged using the IVIS spectrophotometer of PerkinElmer (Waltham, MA, USA) under the following conditions: Level-Hish, Em-840, Ex-745, Epi-illumination, Bin: (30) 8, FOV: 13.2, £2.0.63, Living Image Version: 4.7.4.21053 (14 August 2020), Camera: IS2105N8094, Andor, iKon. The acquired images can be used for noise removal and calculation of fluorescence signal intensity in the heart and liver using IVIS software (v4.4). The data were statistically analyzed by calculating the ratio of the fluorescence intensity of the heart to that of the liver, known as adaptive normalization.

2.10. Statistical Analysis

Mean ± standard deviation (SD) expresses the data. Paired t-tests were conducted within each group to make comparisons. One-way analysis of variance (ANOVA) was used for comparisons between groups, and multiple comparisons or t-tests were used as appropriate. $p < 0.05$ is a significant difference.

3. Results

3.1. HSA-C and CI's Synthesis and Characterization

To verify the targeting of the CI probe, we performed a series of experiments. Firstly, we used western blotting (Figure 1A), and the results clearly showed that the expression level of TfR1 was significantly increased in the MI mouse model compared with the Sham group, which was statistically significant (Figure S1). This result demonstrates that the TfR1-targeting ligand can accurately recognize the target in the MI mouse model. To further verify this conclusion, we performed immunohistochemical staining (Figure S2) which showed that TfR1 was highly expressed at the site of myocardial infarction. We can intuitively observe that HSA-C has no obvious effect on tissues in MI mice, as seen in Figure S3. This result indicated that HSA-C did not cause significant tissue damage under the experimental conditions. Subsequently, we performed Masson staining experiments (Figure 1B). The staining results showed that there was a large amount of fibrous tissue in the MI model compared with the Sham group, indicating a higher degree of fibrosis. This is consistent with the experimental appearance and histopathological changes observed in the MI model, further validating our conjecture. We synthesized HSA-C that can degrade fibrosis, as well as CI, an optical probe targeting TfR1, and evaluated their biodistribution in vivo. The hydrated HSA-C particle sizes are shown in Figure 1C. After CRT peptide labeling, the UV-VIS absorption spectrum of CI changed and was enhanced at approximately 200 nm (Figure 1D). In addition, the binding of ICG to CRT did not alter its fluorescence properties; the fluorescence spectra of CON and CI are shown in Figure 1E. The NIR properties of different CI concentrations are of great significance in the detection of TfR1 in vivo. The NIR fluorescence spectral intensity ($y = 1.57 \times 10^6 x + 1.193 \times 10^7$, R2 = 0.9817, $p < 0.01$) and the concentration of CI were linearly dependent (Figure 1F). This finding suggests that by monitoring the concentration changes of CI, we can indirectly understand the expression level of TfR1. In in vivo detection, this correlation has important implications for diagnosis and treatment monitoring of MI-induced cardiac injury. In addition, different concentrations of CI had no significant effect on H9c2 cell viability (Figure 1G).

3.2. *NIR Fluorescence Imaging in the MI Mouse Model*

In the mouse model of MI, significant differences were observed between the HSA-C-pretreated group and the untreated group. Masson staining showed that myocardial fibrosis was significantly reduced in the MI group after HSA-C pretreatment, indicating that HSA-C had a significant anti-fibrosis effect (Figure 2A). To further quantify this effect, we employed NIR fluorescence imaging for real-time monitoring. The data showed that after HSA-C pretreatment, the target probe at the cardiac site showed the highest fluorescence intensity at 48 h after CI injection into the tail vein. To explore this phenomenon in depth, we conducted a comparative analysis of the groups that received different treatments (Figures 2B and S4). Compared with the MI + CON group, the normalized fluorescence intensity of the MI + CI group was significantly enhanced 1.28-fold. Compared with the untreated MI + CI group, the fluorescence intensity of the HSA-C pretreated MI + CI group was 1.21 times as much as that of the MI + CI group, which was statistically significant (Figure 2C). To evaluate the biodistribution of CI in vivo, we performed ex vivo fluorogram imaging, see Figure S5, and the results showed that CI signals were enriched in cardiac sites of the MI +HSA-C+CI. These data provide strong evidence that HSA-C pretreatment has a significant antifibrotic effect in the mouse model of MI and is effective in improving the normalized fluorescence intensity and distribution of the targeted probe CI at the site of the injured heart. This finding demonstrates the potential application of NIR fluorescence imaging technology in the evaluation of drug efficacy.

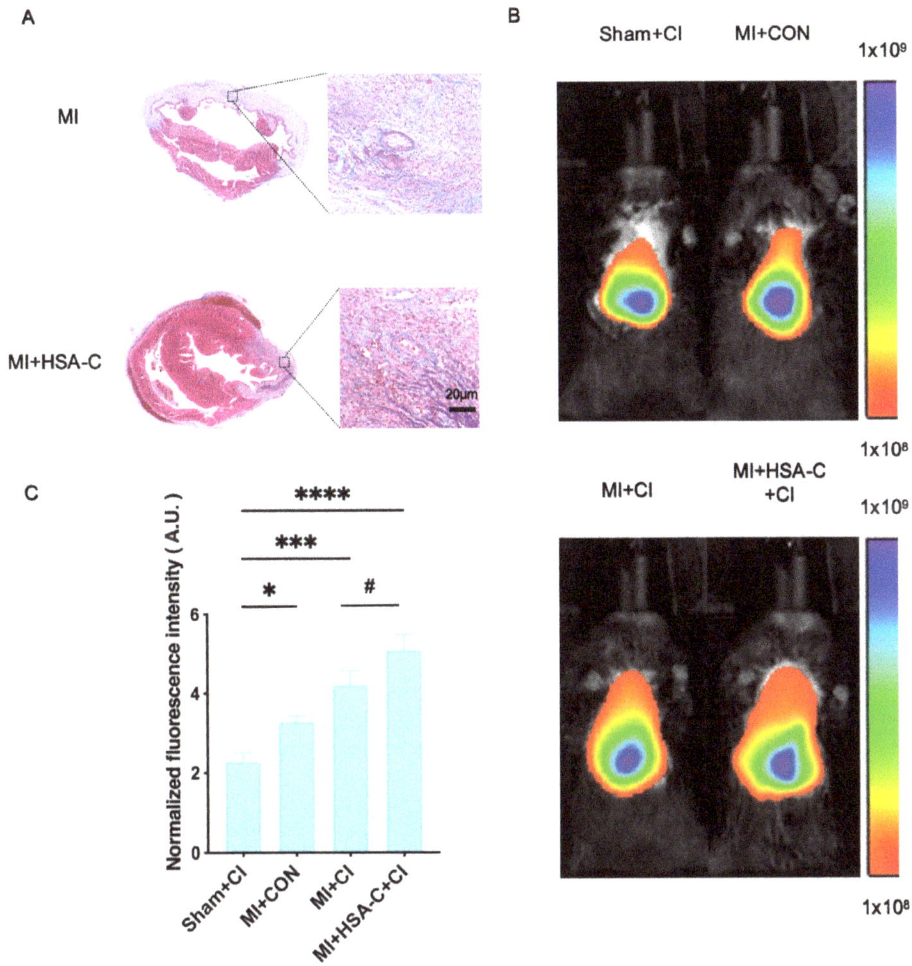

Figure 2. NIR fluorescent imaging of the MI mouse model in vivo. (**A**) The Masson staining images of the MI mouse model and HSA-C preprocessing of the MI mouse model (MI vs. MI + HSA-C; Scale bar = 20 μm). (**B**) The NIR fluorescence images of the Sham + CI, MI + CON, MI + CI, and MI + HSA-C + CI groups at 48 h post-injection in vivo (Sham + CI vs. MI + CON vs. MI + CI vs. MI + HSA-C + CI; n = 3). (**C**) Quantitative comparison of relative fluorescence intensities of Sham + CI/MI + CON/MI + CI/MI + HSA-C + CI (n = 3; one-way ANOVA, # $p < 0.05$; * $p < 0.05$, *** $p < 0.001$, and **** $p < 0.0001$).

4. Discussion

Cardiac fibrosis is a common pathological process that negatively affects the structure and function of the heart. In the process of cardiac fibrosis, cardiomyocytes are gradually replaced by fibrous tissue which leads to a decrease in the systolic and diastolic function of the heart. In addition, cardiac fibrosis may also affect the sensitivity and specificity of molecular imaging. Due to the presence of fibrous tissue, some molecular imaging agents may not bind to cardiomyocytes normally, or the probes may be unspecifically deposited at the site of fibrosis, resulting in inaccurate imaging results. HSA-C can effectively improve cardiac fibrosis by inhibiting the synthesis of collagen, thereby reducing the formation of fibrous tissue. In the animal experiment, HSA-C could significantly reduce the degree of cardiac fibrosis. In addition, HSA-C can also improve the sensitivity and specificity of molecular imaging. Due to the reduction of fibrous tissue, injured cardiomyocytes can be better combined with specific probes to reduce nonspecific deposition of untargeted probes, thereby improving the accuracy of imaging results.

In recent years, ferroptosis, a new disease mechanism, has gradually become a research hotspot. This particular mode of cell death is not only closely related to common diseases such as cardiovascular diseases, tumors, and neurodegenerative diseases, but also involved in multiple other fields [9,19–21]. Several studies have demonstrated a close relationship between ferroptosis and cardiac ischemic injury. Due to the distinctive functions of excessive iron and oxidative damage to lipids in causing harm to the heart muscles, ferroptosis has emerged as both a potential etiology and a therapeutic strategy for myocardial injury associated with ischemic diseases [22]. Abnormal TfR1 expression and the ameliorative effect of iron homeostasis therapy on cardiac injury are associated with abnormal iron metabolism in multiple clinical trials [23–25]. The overexpression and alteration of iron transport play key roles in developing ferroptosis-related cardiac lesions. Therefore, we hypothesized that molecular imaging targeting TfR1 may be useful for accurately detecting MI injury. At present, there are few reports on molecular imaging in vivo detection methods for ferroptosis, which weakens the rationality and enthusiasm of developing drugs for the treatment of ferroptosis [26–28]. Therefore, the present study further proposes that TfR1 overexpression can be used as a biomarker of ferroptosis considering iron homeostasis in healthy participants. To explore biomarkers of ferroptosis more deeply, we introduced near-infrared fluorescence imaging. This technique has the advantages of high sensitivity and absence of ionizing radiation, which provide an ideal means for in vivo imaging. In this study, we designed a novel NIR fluorescence imaging strategy based on HSA-C degradation of myocardial fibrosis combined with ferroptosis detection. This strategy provides three important advantages for imaging: first, it reduces the uptake of nonspecific probes that injure cardiomyocytes and improves the sensitivity and specificity of the signal. Secondly, HSA-C can help to improve the condition of myocardial fibrosis, thus providing a new idea for treatment. In addition, we conducted experiments and data analysis to verify the validity and reliability of this strategy. By comparing with CI alone, we found that the HSA-C pretreatment + CI strategy increased the normalized signal intensity 1.21-fold, showing its superiority in improving the imaging effect.

Previous studies have described methods for visualizing ferroptosis using PET/CT or MRI [29–31]. To our knowledge, this is the first study to use HSA-C and CI to prepare a green, safe, material-specific, and simple method for synthesizing myocardial injury imaging probes in vivo. The HSA-C pretreated CI probe strategy can detect abnormalities in intracellular TfR1 levels following MI-induced cardiac injuries. This method offers a robust approach to circumvent the impact of cardiac remodeling and assess cardiac lesions. It enables real-time evaluation of cardiac injuries following MI, facilitating timely adjustment of treatment plans and enhancing therapeutic outcomes. Although we have made some progress in the study of myocardial injury in MI, there are still some limitations and challenges. First, our study has so far only been validated in a mouse model, which warrants further validation in large animal models and humans. This is because although the mouse model can provide us with a certain reference, there are significant differences from humans in physiology and metabolism. Therefore, more extensive and in-depth studies are needed to ensure the validity of our findings in humans. The HSA-C pretreatment CI probe strategy that we currently use may not apply to all models of myocardial injury. This is because different types of myocardial injury may have different mechanisms and processes, so appropriate methods and techniques need to be selected for specific situations. To address this issue, we need to further optimize and improve the existing probe strategy so that it can be better applied to various models of myocardial injury. In addition, further monitoring is also needed to target the possible adverse effects of HSA-C on cardiac injury after MI.

In summary, a novel HSA-C and CI imaging strategy was established to assess MI-induced myocardial injury in vivo. This study has important theoretical significance and practical value, and provides new ideas and methods for the research and treatment of in vivo visualization of myocardial ferroptosis. We will continue to study the mechanism of myocardial injury, in order to bring better detection methods for patients with myocardial injury and improve the treatment effect.

Supplementary Materials: The following supporting information can be downloaded at: https://www.mdpi.com/article/10.3390/biology13040213/s1. Figure S1. The western blot analysis of TfR1 in in MI and Sham group (n = 3, Student's t test. ** $p < 0.01$). Figure S2. A high level of TfR1 expression was detected in the heart tissue of the MI mouse model by immunohistochemical staining. (Scale bar = 50 μm). Figure S3. H&E staining of major organs in mouse which were pretreated with HSA-C. (Scale bar: 50 μm). Figure S4. (A) The NIR fluorescence images of the MI mouse model at 48 h post-injection from different groups (Sham + CON vs. Sham + CI; n = 3). (B) Quantitative comparison of normalized fluorescence intensities of different group (Sham + CON vs. Sham + CI; n = 3). Figure S5. The ex vivo imaging for the heart and the other major organs of the MI mouse model (MI + CON vs. MI + HSA-C + CI), which were removed and captured 48 h after injection, were obtained using NIR fluorescence (H: Heart; Lu: Lung; Li: Liver; S: Spleen; K: Kidney). Figure S6. Figure of Western Blot results: Comparison of TfR1 protein expression between myocardial infarction group and sham operation group.

Author Contributions: Conceptualization: W.Y., Y.W., L.L. and F.L.; Methodology: W.Y., Y.W., Y.P. and L.L.; Investigation: W.Y., Y.W., F.L., A.L. and H.L.; Visualization: W.Y. and Y.W.; Supervision: L.T., L.L., H.Q. and C.F.; Writing—original draft: W.Y. and Y.W.; Writing—review and editing: all authors. All authors have read and agreed to the published version of the manuscript.

Funding: This study was supported by the Science and Technology Innovation Project of China Academy of Chinese Medical Sciences (NO. CI2021A05013).

Institutional Review Board Statement: The animal study protocol was approved by the Ethics Committee of the Institutional Animal Care and Use Committee of Xiyuan Hospital, China Academy of Chinese Medical Sciences (protocol code 2021XLC027 and Nov. 10, 2021 of approval).

Informed Consent Statement: Not applicable.

Data Availability Statement: After careful consideration, we have added research data to the additional information to ensure the completeness and accuracy of the information.

Acknowledgments: The authors would like to acknowledge the instrumental and technical support of Multimodal Biomedical Imaging Experimental Platform, Institute of Automation, Chinese Academy of Sciences.

Conflicts of Interest: The authors declare no conflicts of interest.

References

1. Aghagolzadeh, P.; Plaisance, I.; Bernasconi, R.; Treibel, T.A.; Quetglas, C.P.; Wyss, T.; Wigger, L.; Nemir, M.; Sarre, A.; Chouvardas, P.; et al. Assessment of the Cardiac Noncoding Transcriptome by Single-Cell RNA Sequencing Identifies FIXER, a Conserved Profibrogenic Long Noncoding RNA. *Circulation* **2023**, *148*, 778–797. [CrossRef]
2. Frangogiannis, N.G. Pathophysiology of Myocardial Infarction. *Compr. Physiol.* **2015**, *5*, 1841–1875. [CrossRef]
3. Querejeta, R.; López, B.; González, A.; Sánchez, E.; Larman, M.; Ubago, J.L.M.; Díez, J. Increased collagen type I synthesis in patients with heart failure of hypertensive origin: Relation to myocardial fibrosis. *Circulation* **2004**, *110*, 1263–1268. [CrossRef]
4. Snider, J.C.; Riley, L.A.; Mallory, N.T.; Bersi, M.R.; Umbarkar, P.; Gautam, R.; Zhang, Q.; Mahadevan-Jansen, A.; Hatzopoulos, A.K.; Maroteaux, L.; et al. Targeting 5-HT 2B Receptor Signaling Prevents Border Zone Expansion and Improves Microstructural Remodeling After Myocardial Infarction. *Circulation* **2021**, *143*, 1317–1330. [CrossRef] [PubMed]
5. Bengel, F.M.; Diekmann, J.; Hess, A.; Jerosch-Herold, M. Myocardial Fibrosis: Emerging Target for Cardiac Molecular Imaging and Opportunity for Image-Guided Therapy. *J. Nucl. Med.* **2023**, *64* (Suppl. S2), 49S–58S. [CrossRef] [PubMed]
6. Otto, G. Regenerative medicine: Stem cell-derived cardiomyocytes heal a broken heart. *Nat. Rev. Drug Discov.* **2018**, *17*, 622. [CrossRef]
7. Barton, A.K.; Tzolos, E.; Bing, R.; Singh, T.; Weber, W.; Schwaiger, M.; Varasteh, Z.; Slart, R.H.J.A.; Newby, D.E.; Dweck, M.R. Emerging molecular imaging targets and tools for myocardial fibrosis detection. *Eur. Heart J. Cardiovasc. Imaging* **2023**, *24*, 261–275. [CrossRef]

8. Gupta, S.; Ge, Y.; Singh, A.; Gräni, C.; Kwong, R.Y. Multimodality Imaging Assessment of Myocardial Fibrosis. *JACC Cardiovasc. Imaging* **2021**, *14*, 2457–2469. [CrossRef] [PubMed]
9. Fang, X.; Ardehali, H.; Min, J.; Wang, F. The molecular and metabolic landscape of iron and ferroptosis in cardiovascular disease. *Nat. Rev Cardiol.* **2023**, *20*, 7–23. [CrossRef]
10. Li, N.; Jiang, W.; Wang, W.; Xiong, R.; Wu, X.; Geng, Q. Ferroptosis and its emerging roles in cardiovascular diseases. *Pharmacol. Res.* **2021**, *166*, 105466. [CrossRef]
11. Chen, L.; Ma, Y.; Ma, X.; Liu, L.; Jv, X.; Li, A.; Shen, Q.; Jia, W.; Qu, L.; Shi, L.; et al. TFEB regulates cellular labile iron and prevents ferroptosis in a TfR1-dependent manner. *Free Radic. Biol. Med.* **2023**, *208*, 445–457. [CrossRef] [PubMed]
12. Jiang, X.; Stockwell, B.R.; Conrad, M. Ferroptosis: Mechanisms, biology and role in disease. *Nat. Rev. Mol. Cell Biol.* **2021**, *22*, 266–282. [CrossRef] [PubMed]
13. Cheng, P.; Han, H.; Chen, F.; Cheng, L.; Ma, C.; Huang, H.; Chen, C.; Li, H.; Cai, H.; Huang, H.; et al. Amelioration of acute myocardial infarction injury through targeted ferritin nanocages loaded with an ALKBH5 inhibitor. *Acta Biomater.* **2022**, *140*, 481–491. [CrossRef]
14. SNMMI Image of the Year 2022: PET/CT Biomarker Predicts Post-MI Cardiac Remodeling. *J. Nucl. Med.* **2022**, *63*, 16N.
15. van den Bos, E.J.; Baks, T.; Moelker, A.D.; Kerver, W.; van Geuns, R.-J.; van der Giessen, W.J.; Duncker, D.J.; Wielopolski, P.A. Magnetic resonance imaging of haemorrhage within reperfused myocardial infarcts: Possible interference with iron oxide-labelled cell tracking? *Eur. Heart J.* **2006**, *27*, 1620–1626. [CrossRef]
16. Desai, N.D.; Moussa, F.; Singh, S.K.; Chu, P.; Fremes, S.E. Intraoperative fluorescence angiography to determine the extent of injury after penetrating cardiac trauma. *J. Thorac. Cardiovasc. Surg.* **2008**, *136*, 218–219. [CrossRef] [PubMed]
17. Zhuo, J.; Wang, Y.; Hui, H.; Li, C.; Yang, J.; Zhang, P.; Fang, C.; Tian, J. Enhanced glypican-3-targeted identification of hepatocellular carcinoma with liver fibrosis by pre-degrading excess fibrotic collagen. *Acta Biomater.* **2023**, *158*, 435–448. [CrossRef]
18. Du, Y.; Gao, J.; Zhang, H.; Meng, X.; Qiu, D.; Gao, X.; Zheng, A. Brain-targeting delivery of MMB4 DMS using carrier-free nanomedicine CRT-MMB4@MDZ. *Drug Deliv.* **2021**, *28*, 1822–1835. [CrossRef]
19. Sahebkar, A.; Foroutan, Z.; Katsiki, N.; Jamialahmadi, T.; Mantzoros, C.S. Ferroptosis, a new pathogenetic mechanism in cardiometabolic diseases and cancer: Is there a role for statin therapy? *Metabolism* **2023**, *146*, 155659. [CrossRef]
20. Zhou, R.-P.; Chen, Y.; Wei, X.; Yu, B.; Xiong, Z.-G.; Lu, C.; Hu, W. Novel insights into ferroptosis: Implications for age-related diseases. *Theranostics* **2020**, *10*, 11976–11997. [CrossRef]
21. Ryan, S.K.; Ugalde, C.L.; Rolland, A.-S.; Skidmore, J.; Devos, D.; Hammond, T.R. Therapeutic inhibition of ferroptosis in neurodegenerative disease. *Trends Pharmacol. Sci.* **2023**, *44*, 674–688. [CrossRef]
22. Lillo-Moya, J.; Rojas-Solé, C.; Muñoz-Salamanca, D.; Panieri, E.; Saso, L.; Rodrigo, R. Targeting Ferroptosis against Ischemia/Reperfusion Cardiac Injury. *Antioxidants* **2021**, *10*, 667. [CrossRef] [PubMed]
23. Baune, B.T.; Neuhauser, H.; Ellert, U.; Berger, K. The role of the inflammatory markers ferritin, transferrin and fibrinogen in the relationship between major depression and cardiovascular disorders—The German Health Interview and Examination Survey. *Acta Psychiatr. Scand.* **2010**, *121*, 135–142. [CrossRef] [PubMed]
24. Stack, A.G.; Mutwali, A.I.; Nguyen, H.T.; Cronin, C.; Casserly, L.; Ferguson, J. Transferrin saturation ratio and risk of total and cardiovascular mortality in the general population. *QJM* **2014**, *107*, 623–633. [CrossRef]
25. Nikkari, S.T.; Koivu, T.A.; Anttila, P.; Raunio, I.; Sillanaukee, P. Carbohydrate-deficient transferrin and gamma-glutamyltransferase are inversely associated with lipid markers of cardiovascular risk. *Eur. J. Clin. Investig.* **1998**, *28*, 793–797. [CrossRef] [PubMed]
26. Zhao, D.; Huang, X.; Tian, Y.; Zou, J.; Wang, F.; Chen, X. Fluorescence Imaging-Incorporated Transcriptome Study of Glutathione Depletion-Enhanced Ferroptosis Therapy via Targeting Gold Nanoclusters. *ACS Appl. Mater. Interfaces* **2023**, *15*, 6385–6396. [CrossRef] [PubMed]
27. Wu, G.L.; Liu, F.; Li, N.; Fu, Q.; Wang, C.; Yang, S.; Xiao, H.; Tang, L.; Wang, F.; Zhou, W.; et al. Trisulfide Bond-Mediated Molecular Phototheranostic Platform for "Activatable" NIR-II Imaging-Guided Enhanced Gas/Chemo-Hypothermal Photothermal Therapy. *Adv. Sci.* **2023**, *10*, e2304104. [CrossRef] [PubMed]
28. Yang, W.; Wang, Y.; Fu, C.; Li, C.; Feng, F.; Li, H.; Tan, L.; Qu, H.; Hui, H.; Wang, J.; et al. Quantitative visualization of myocardial ischemia-reperfusion-induced cardiac lesions via ferroptosis magnetic particle imaging. *Theranostics* **2024**, *14*, 1081–1097. [CrossRef] [PubMed]
29. Yang, X.; Zhang, S.; Lai, M.; Ji, X.; Ye, Y.; Tang, J.; Liu, X.; Zhao, M. Fluorescent probes for lighting up ferroptotic cell death: A review. *Talanta* **2023**, *260*, 124628. [CrossRef]
30. Zhou, R.; Liu, Y.; Wang, Z.; Lv, J.; Liao, W.; Shen, Z.; Rong, X. Nanoparticle-Based MRI-Guided Tumor Microenvironment Heating via the Synergistic Effect of Ferroptosis and Inhibition of TGF-β Signaling. *Adv. Healthc. Mater.* **2023**, *12*, e2300176. [CrossRef]
31. Zeng, F.; Nijiati, S.; Liu, Y.; Yang, Q.; Liu, X.; Zhang, Q.; Chen, S.; Su, A.; Xiong, H.; Shi, C.; et al. Ferroptosis MRI for early detection of anticancer drug-induced acute cardiac/kidney injuries. *Sci. Adv.* **2023**, *9*, eadd8539. [CrossRef] [PubMed]

biology

MDPI

Article

Gold-Nanorod-Assisted Live Cell Nuclear Imaging Based on Near-Infrared II Dark-Field Microscopy

Yifeng Shi [1,†], Shiyi Peng [2,†], Zhongyu Huang [1], Zhe Feng [2], Wen Liu [1,*], Jun Qian [2,*] and Weidong Zhou [1]

1 Key Laboratory of Optical Information Detection and Display Technology of Zhejiang, Zhejiang Normal University, Jinhua 321004, China; shiyifeng@zjnu.edu.cn (Y.S.); hzhongyu@zjnu.edu.cn (Z.H.); wdzhou@zjnu.cn (W.Z.)

2 State Key Laboratory of Modern Optical Instrumentations, Centre for Optical and Electromagnetic Research, College of Optical Science and Engineering, International Research Center for Advanced Photonics, Zhejiang University, Hangzhou 310058, China; 12330003@zju.edu.cn (S.P.); zhefeng@zju.edu.cn (Z.F.)

* Correspondence: wenliu@zjnu.edu.cn (W.L.); qianjun@zju.edu.cn (J.Q.)

† These authors contributed equally to this work.

Simple Summary: Colorectal cancer cells exhibited superior dark-field imaging results in the near-infrared II (NIR-II) wavelength range (900–1880 nm) compared to the visible light region. Subsequently, we synthesized gold nanorods (GNRs) for dark-field scattering imaging of colorectal cancer cells in the NIR-II spectrum. The results demonstrated that imaging with GNRs significantly improved the signal-to-background ratio (SBR) and showed enhanced performance, particularly in the 1425–1475 nm wavelength range. Finally, we conducted dark-field imaging of cell nuclei in the NIR-II range utilizing GNRs for specific labeling of colorectal cancer cell nuclei. The resulting nuclear images were highly accurate in localization and exhibited higher SBR compared to non-specifically-labeled cell imaging.

Abstract: Dark-field microscopy offers several advantages, including high image contrast, minimal cell damage, and the absence of photobleaching of nanoprobes, which make it highly advantageous for cell imaging. The NIR-II window has emerged as a prominent research focus in optical imaging in recent years, with its low autofluorescence background in biological samples and high imaging SBR. In this study, we initially compared dark-field imaging results of colorectal cancer cells in both visible and NIR-II wavelengths, confirming the superior performance of NIR-II imaging. Subsequently, we synthesized gold nanorods with localized surface plasmon resonance (LSPR) absorption peaks in the NIR-II window. After bio-compatible modification, we non-specifically labeled colorectal cancer cells for NIR-II dark-field scattering imaging. The imaging results revealed a sixfold increase in SBR, especially in the 1425–1475 nm wavelength range. Finally, we applied this imaging system to perform dark-field imaging of cell nuclei in the NIR-II region and used GNRs for specific nuclear labeling in colorectal cancer cells. The resulting images exhibited higher SBR than non-specifically-labeled cell imaging, and the probe's labeling was precise, confirming the potential application of this system in photothermal therapy and drug delivery for cancer cells.

Keywords: dark-field microscopy; gold nanorods; near-infrared second window; live cell nuclear imaging

1. Introduction

Biological imaging is a crucial component of modern medicine. Optical imaging techniques offer advantages, such as radiation-free imaging, high spatial and temporal resolution, low photon damage, and real-time imaging capabilities, when compared to traditional imaging methods. Therefore, optical biological imaging techniques have unique advantages in many applications. The structure and function of biological organisms are fundamentally based on cells, particularly the cell nucleus. Various optical microscopy techniques are commonly used for cellular imaging, including bright-field imaging, dark-field imaging, and fluorescence imaging [1,2]. Among these, bright-field imaging is the most basic microscopy technique, but it often exhibits low image contrast, making it more suitable for samples with strong absorption. As research has advanced in imaging systems, more sophisticated microscopy techniques, such as fluorescence microscopy, have emerged as essential tools in modern biological research. Fluorescence microscopy [3,4] is highly sensitive and non-invasive, offering high spatial and temporal resolution for imaging live cells and monitoring dynamic cellular processes [5,6]. However, when using fluorescence microscopy, higher laser excitation power is often required, and common dyes used during imaging are susceptible to photobleaching; additionally, the resolution can be limited by the autofluorescence of biological samples. Therefore, for cellular imaging, bright-field and dark-field microscopy are commonly employed, with dark-field microscopy being particularly advantageous. Dark-field microscopy (DFM) [7] is a specialized optical microscopy technique that can replace more complex or expensive microscopy methods in certain applications. DFM allows rapid identification of nanoscale particles, microscale cells, and tissues [8–10] in biological samples and environmental samples, such as wastewater, soil, sediments, etc. [11,12] and provides an effective means for rapid, efficient, cost-effective, and non-destructive imaging of nanoscale particles. DFM illuminates the sample with oblique light, typically using an illumination aperture outside the NA of the objective lens. A central stop or mask is added to the system to block the central light beam from the light source, causing the incident light to be angled onto the sample. Consequently, DFM imaging yields a dark field of view with higher image contrast. It is important to note that DFM does not inherently enhance image resolution but instead improves the contrast of image edges that are obscured by bright-field illumination.

Typically, light in the range of 700–2500 nm is referred to as near-infrared (NIR) light. NIR light is characterized by low scattering in biological tissues and low autofluorescence, making it suitable for biological imaging and yielding images with a higher signal-to-background ratio (SBR) [13,14]. NIR light can be further divided into two windows: the first near-infrared window (NIR-I, 760–900 nm) and the second near-infrared window (NIR-II, 900–1880 nm). Among these, the NIR-II window has two notable characteristics: first, it exhibits significant water absorption, which reduces background noise from scattered signal photons in wide-field imaging techniques; second, it offers low scattering, enabling deeper imaging within living organisms [15]. As such, the NIR-II window provides advantages in terms of imaging depth and image SBR. Hong et al. [16] investigated the absorption spectrum of water at various wavelengths and found that water exhibited absorption peaks at 970 nm, 1200 nm, 1450 nm, and beyond 1800 nm. Among these peaks, the absorption at 1450 nm was particularly strong, significantly affecting imaging quality. However, recent research has shown that the absorption characteristics of this wavelength range can be utilized to improve imaging quality. Feng et al. [17] simulate the photon propagation in the NIR region, confirming the positive contribution of moderate light absorption by water in intravital imaging.

Nanomaterials, owing to their diverse types, wide spectral coverage, chemical stability, and resistance to photobleaching, are frequently employed in biological imaging. Researchers are actively engaged in the development and preparation of NIR-II nanoprobes to further enhance imaging quality. The nanomaterials commonly used can be divided into inorganic materials, such as gold nanorods, carbon nanotube [18], quantum dot [19], rare-earth-doped nanoparticles [20], etc., and organic materials, mainly conjugated polymer [21]. Among these, gold nanorods [22] are rod-shaped nanostructures, with dimensions typically below a few hundred nanometers, composed primarily of gold. GNRs offer stability in chemical properties, tunable scattering peaks [23], diverse synthesis and modification methods [24], flexible size control, and easy conjugation with other molecules to alter their physical and chemical properties, making them extensively researched and applied. In particular, GNRs exhibit unique LSPR properties, resulting from the interaction between electrons and electromagnetic waves at the nanoscale. When the wavelength of incident light is approximately ten times the size of the nanorods, the surface electrons undergo collective oscillations, causing most of the incident photons with frequencies matching the collective oscillation frequency of valence electrons to be absorbed and scattered. The presence of LSPR peaks in the extinction spectrum is a prominent feature at the macroscopic level. These LSPR peaks indicate enhanced absorption and scattering properties of GNRs, making them advantageous for various applications, particularly in the field of biological imaging [25]. GNRs possess a substantial scattering cross-section, and variations in their aspect ratios result in different longitudinal LSPR absorption peaks, enabling them to scatter light at different wavelengths [26]. When GNRs scatter light in the near-infrared range, they offer significant advantages for biological imaging. The enhanced scattering capability of GNRs can be harnessed for dark-field microscopy of cell nuclei.

In this study, taking advantage of the NIR-II window, we optimized a dark-field imaging system and employed GNRs as probes for dark-field imaging of cell nuclei. Initially, we observed that compared to the visible light spectrum, NIR-II imaging offered reduced background noise, clearer cell contours, and increased contrast, particularly with increasing wavelength. However, cell scattering in the NIR-II range was relatively low, resulting in lower light signal intensity. To further enhance imaging quality, we introduced gold nanoprobes to enhance cell scattering and, consequently, signal intensity. We synthesized GNRs using a one-pot method, applied biocompatible polymer modification, and non-specifically labeled live cells for use in NIR-II dark-field scattering imaging. However, polymer-modified GNRs could only enter the cell cytoplasm and could not access the cell nucleus. To image cell nuclei, we modified GNRs with membrane-penetrating proteins and nuclear localization proteins for specific nuclear labeling, ultimately obtaining well-defined nuclear boundary images. The specificity of the probe's labeling closely matched that of nuclear-specific dyes. This demonstrates that our research holds practical significance and offers potential applications in the treatment of cancer cells and drug delivery within the realm of biomedicine.

2. Materials and Methods

2.1. Experimental Reagents

The reagents for GNR preparation were chloroauric acid ($HAuCl_4$), silver nitrate ($AgNO_3$), cetyltrimethylammonium bromide (CTAB), Hydroquinone (HQ), and sodium-borohydride ($NaBH_4$), which were all purchased from Sigma-Aldrich (St. Louis, MO, USA).

The reagents for polymer electrolyte multilayer coating of GNRs were Poly(3,4-ethylenedioxythiophene)-polystyrene sulfonate (PSS, MW = 240,000) purchased from Polysciences, Inc. (Warrington, PA, USA); Poly (allylamine hydrochloride) (PAH) purchased from Beijing Huaweiruike Technology Co., Ltd. (Beijing, China); mPEG-SH (Methoxypoly (ethylene glycol)-thiol, MW = 2000) purchased from Beijing Jiankai Technology Co., Ltd. (Beijing, China); and Sodium chloride (NaCl) purchased from Sinopharm Group Chemical Reagent Co., Ltd. (Shanghai, China).

2.2. NIR-II Extinction Spectroscopy Measurement System

As shown in Figure 1, a mercury lamp was used as the light source; after fixing the sample in a cuvette, the GNRs' extinction spectra in the ranges of 400–900 nm and 900–1600 nm were measured separately using the PG2000 (PG2000PRO, Ideaoptics, Shanghai, China) and NIR2200 (NIR22px, Ideaoptics, Shanghai, China) spectrophotometers, respectively. The acquired data were subsequently presented on a computer.

2.3. NIR-II Dark-Field Microscopic Imaging System

As shown in Figure 2, a dark-field microscope (Eclipse LV150N, Nikon, Tokyo, Japan) was employed in the experiment with a halogen lamp as the light source and an InGaAs camera (SD640, TEKWIN SYSTEM, Shanghai, China) as the photon detector. The light emitted from the lamp passed through an adjustable aperture and a pre-focusing adjustable field stop before achieving Köhler illumination. The imaging system incorporated a dark-field condenser with a dark-field stop, which was a circular shading plate capable of obstructing the central portion of the light beam, allowing only the incident light from the periphery of the circular ring to pass through. After passing through the condenser, the light was inclined from various angles to illuminate the sample surface. A low numerical aperture objective (50×, NA0.8, air) was used in the system, ensuring minimal collection of incident light and primarily collecting the scattered light from the sample. The collected light signal was then used to generate images on a computer.

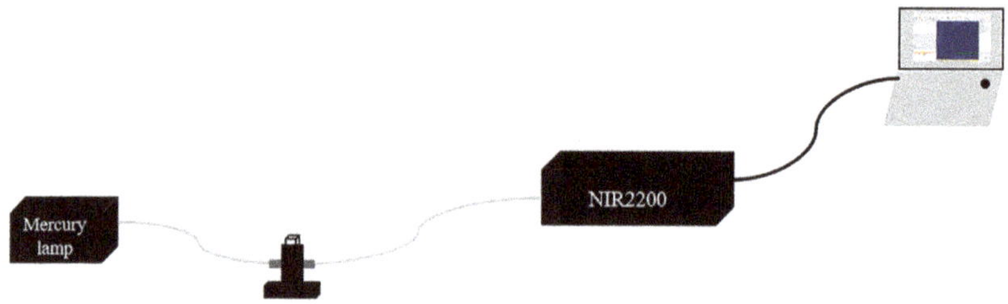

Figure 1. Extinction spectrum test system. A mercury lamp was used as the light source; PG2000 and NIR2200 spectrometers were employed to measure the extinction spectra.

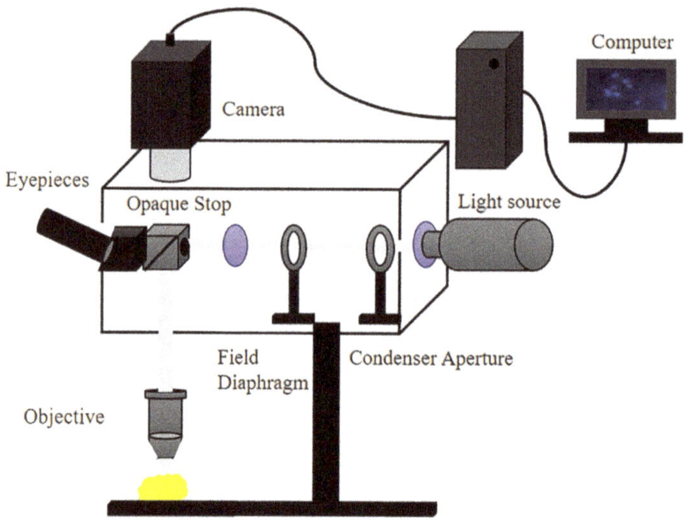

Figure 2. Dark-field microscopy imaging system. The system was sensitive to the 900–1700 nm wavelength range by replacing the commercial visible CCD with an InGaAs camera. In this experiment, different long-pass filters were employed to capture cell images at various wavelengths within this range.

2.4. Calculation of SBR

In order to calculate the intensity of the signal portion in the image, we selected five different regions and obtained the average grayscale values, resulting in IS. The average background intensity IB was similarly obtained using the same method.

The SBR was calculated using the following formula:

$$\mathrm{SBR} = \frac{\mathrm{I_S}}{\mathrm{I_B}}$$

2.5. One-Pot Synthesis of GNRs

At room temperature, under gentle magnetic stirring, 9.6 mL of deionized water was added to a flask, followed by the addition of $HAuCl_4$ solution (0.4 mL, 25 mM). Afterward, a CTAB solution (10 mL, 0.2 mM) was added to the mixture, which turned from pale yellow to orange. After that, $AgNO_3$ solution (0.1 mL, 1 M) and HQ solution (1 mL, 0.05 M) were added sequentially. $AgNO_3$ facilitated the longitudinal growth of gold nanorods, while HQ reduced Au^{3+} to Au^{+}, gradually rendering the solution colorless. When the solution was completely colorless, an ice-cold $NaBH_4$ solution (10 μL, 10 mM) was added with vigorous stirring. After one minute, the mixture was placed in a constant temperature drying oven at 30 °C for 24 h to grow the gold nanorods. The solution changed from colorless to dark brown. Excess reactants were removed through centrifugation (9600 rpm, 12 min), yielding GNRs [27].

2.6. Polymer Electrolyte Multilayer Coating of GNRs

To remove excess CTAB from the GNRs, the water dispersion of GNRs (10 mL) was centrifuged and washed with water twice. The GNRs were then redispersed in 5 mL of deionized water. PSS solution (30 mM) and NaCl solution (30 mM) were each added (0.5 mL) to the GNR water dispersion. The mixture was vigorously oscillated for 3 min using a vortex oscillator and then stirred for an additional 3 hours. Excess PSS was removed through centrifugation (7200 rpm, 10 min) and washed with water. The resulting product was redispersed in 5 mL of water, yielding GNR-PSS water dispersion. Because the surface of GNRs after PSS characterization is negatively charged and the cell surface is also negatively charged, GNRs need to be modified using PAH. We used a concentration of 10 mM NaCl as a solvent to configure a concentration of 10 mg/mL PAH solution, and we also added 0.5 mL each to the GNR-PSS water dispersion. PAH multilayer modification of GNRs was performed similarly. Finally, the modified GNR-PSS-PAH can be directly incubated with the cells and enter the cells under the action of electrostatic adsorption for subsequent experiments [28].

2.7. Preparation of GNR-Based Nuclear Targeting Probes

After centrifugation and washing, m-PEG-SH (400 μL, 10 mg/mL, MW = 2000), membrane localization protein (RGD) (6 μL, 5 mM), and nuclear localization protein (NLS) (25 μL, 5 mM) were added to GNRs (10 mL, 0.1 nM) synthesized using the one-pot method. The mixture was stirred magnetically for 24 h at room temperature. Excess reactants were removed through centrifugation (6000 rpm, 8 min), and the resulting product was redispersed in 1 mL of water.

2.8. Characterization of GNRs

The morphologies of GNRs were also studied using a transmission electron microscope (TEM, JEM-1200EX, JEOL, Tokyo, Japan) operating at 80 kV in bright-field mode. The hydrodynamic size distribution was determined through dynamic light scattering (DLS) with a particle size analyzer (Malvern, Zetasizer 3000 HAS, Malvern, UK) at a fixed angle of 90° at room temperature. The zeta potential was determined with the same analyzer (Malvern, Zetasizer 3000 HAS).

2.9. Cell Culturing

CT26 cells were used in the experiment and cultured in DMEM supplemented with 10% fetal bovine serum, 1% penicillin–streptomycin, and 1% amphotericin B. Cells were maintained at 37 °C in a humidified atmosphere containing 95% air and 5% carbon dioxide. CT26 cells were seeded in a 24-well plate and incubated in a cell culture incubator for 24 h before sample addition. We used samples of GNR-PSS-PAH and GNR-RGD-NLS with a concentration of 1 nM and added 40 μL to two petri dishes, respectively, and then cultured them with cells for 24 h. We also set up a blank control group.

3. Results

3.1. Dark-Field Scattering Imaging of Colon Cancer Cells

In our experiments, CT26 cells, a type of colon cancer cell line, were subjected to dark-field scattering imaging in both the visible light spectrum and the NIR-II spectrum (900–1700 nm). As shown in Figure 3a,b, compared to imaging results in the visible spectrum, the imaging quality in the NIR-II spectrum was superior, allowing for clearer visualization of the cell morphology. Upon calculating the SBR for both spectra (Figure 3c), the SBR for the visible spectrum image was 1.3, whereas the SBR for the NIR-II image was 3.4, indicating a higher SBR for the NIR-II image. This confirmed that dark-field imaging of cells in the NIR-II spectrum was more effective. We hypothesized that the improved performance in the NIR-II spectrum was due to the superior anti-scattering ability of longer wavelengths, resulting in fewer signal photons being scattered in the background. Additionally, the visible spectrum is more susceptible to environmental light noise, resulting in a more pronounced background. Additionally, the visible spectrum is more prone to environmental light noise, making background signals more prominent.

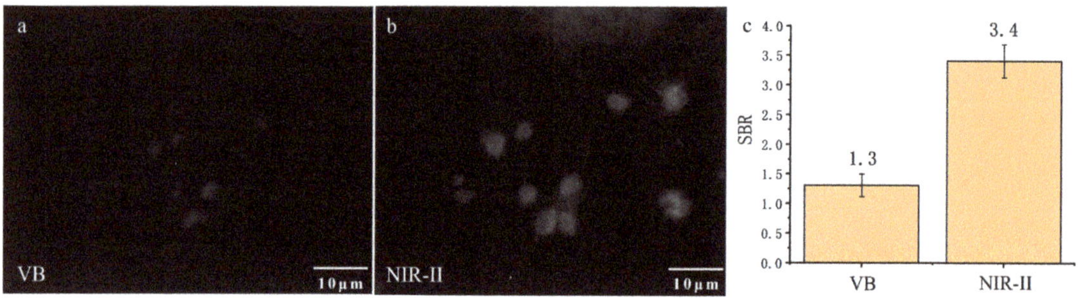

Figure 3. Comparison of imaging effect between visible band and NIR-II band: (**a**) visible band; (**b**) NIR-II window (900–1700 nm); (**c**) shows SBR of the two images.

3.2. Dark-Field Scattering Imaging of Colon Cancer Cells in the NIR-II Window

To further investigate the impact of wavelength on imaging, we introduced long-pass filters (LP) ranging from 900 to 1500 nm and a band-pass filter (BP) at 1450 nm into the system, thereby capturing cell images in the corresponding wavelength bands (Figure 4a–h). SBR analysis of the images was performed (Figure 4i). As the imaging wavelength shifted towards longer wavelengths, the SBR gradually increased for bands from 900 to 1700 nm, 1000 to 1700 nm, 1100 to 1700 nm, and 1200 to 1700 nm. However, for bands from 1300 to 1700 nm to 1500 to 1700 nm, SBR exhibited a decreasing trend with the wavelength shift. The cross-sectional signal along the red dashed line of the cell in Figure 4d is presented in Figure S1, indicating a relatively uniform intensity of the cytoplasm and nucleus.

From these results, we observed that as the imaging wavelength shifted towards longer wavelengths, the overall imaging quality improved. However, after the 1200–1700 nm range, the imaging quality began to decline. The reason was speculated to be that as the wavelength increased, the cell's scattering decreased due to the limited cell size. There is a competition between signal photons and background photons. The decreased scattering resulted in weaker background photons, which improved the imaging quality. Meanwhile, decreased signal photons were harmful to the imaging quality. To leverage the advantages of imaging in the NIR-II window more effectively, we introduced gold nanorods as contrast agents to enhance cell scattering of longer-wavelength light.

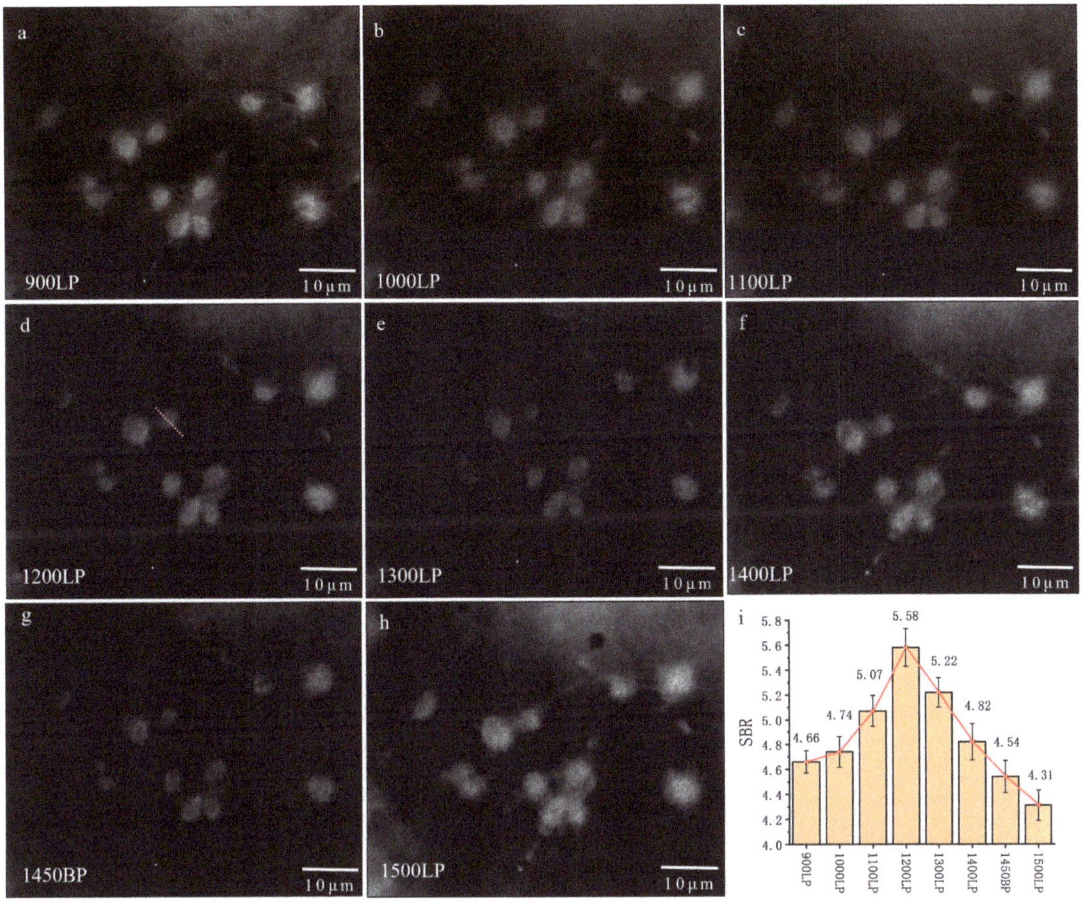

Figure 4. Dark-field imaging of cells in NIR-II window. We added (**a**) 900 LP, (**b**) 1000 LP, (**c**) 1100 LP, (**d**) 1200 LP, (**e**) 1300 LP, (**f**) 1400 LP, (**g**) 1450 BP, and (**h**) 1500 LP to the system; (**i**) shows the comparison of SBR of the images above.

3.3. NIR-II Dark-Field Scattering Imaging of Cells Assisted by GNRs

We synthesized gold nanorods through a one-pot method and performed multilayer modification of the nanorods using polymer electrolytes to enhance their biocompatibility (Figure 5a) [29]. The GNRs synthesized had a longitudinal length of 92.2 $\pm$ 8.1 nm and an axial length of 10.2 $\pm$ 0.8 nm according to the TEM image, as shown in Figure 5d, and the aspect ratio was about 9.0. The high aspect ratio allowed a redshifted LSPR peak at 1044 nm, which would provide a strong scattering in the NIR-II region. After synthesizing the gold nanorods, we successfully confirmed that they did not aggregate after multiple rounds of centrifugation and washing, indicating a successful coating process. The size of the gold nanorods changed after coating, affecting their localized surface plasmon resonance (LSPR) and resulting in a redshift in the LSPR peak (Figure 5b). Specifically, after multiple layers of coating, the LSPR peak shifted from 1044 nm to 1098 nm, indicating successful PSS and PAH coatings. It can be noticed that two peaks at about 900 nm and 300 nm were not supposed to appear, and they were actually due to the experimental setup and data analysis. Additionally, the zeta potential of GNRs in water dispersion was measured to be +57.3 mV, which changed to −21.5 mV after PSS coating and further increased to +35.7 mV after PAH coating (Figure 5c). These data collectively confirm the successful polymer electrolyte coating on the GNRs' surfaces. The average hydrodynamic diameter of the GNRs was 16.6 ± 4.6 nm, and that of GNR-PSS-PAH was 75.2 ± 22.3 nm, as shown in Figure 5d,e, respectively. The large difference was because the long chains of polymers would stretch out in water. Considering the measurement model of hydrodynamic diameters was spherical, when gold nanorods with a high aspect ratio were measured, the diameter of equivalent spheres would be much smaller than the real longitudinal length in TEM images. The polymer dispersity index (PDI) of the GNRs and GNR-PSS-PAH was 0.08 and 0.09, which indicated a relatively uniform molecular weight distribution. The TEM size of GNR-PSS-PAH was much smaller than the hydrodynamic diameter because of the drying process for TEM measurement and the low contrast of PSS-PAH in TEM images. All of these results further proved the successful coating of PSS and PAH.

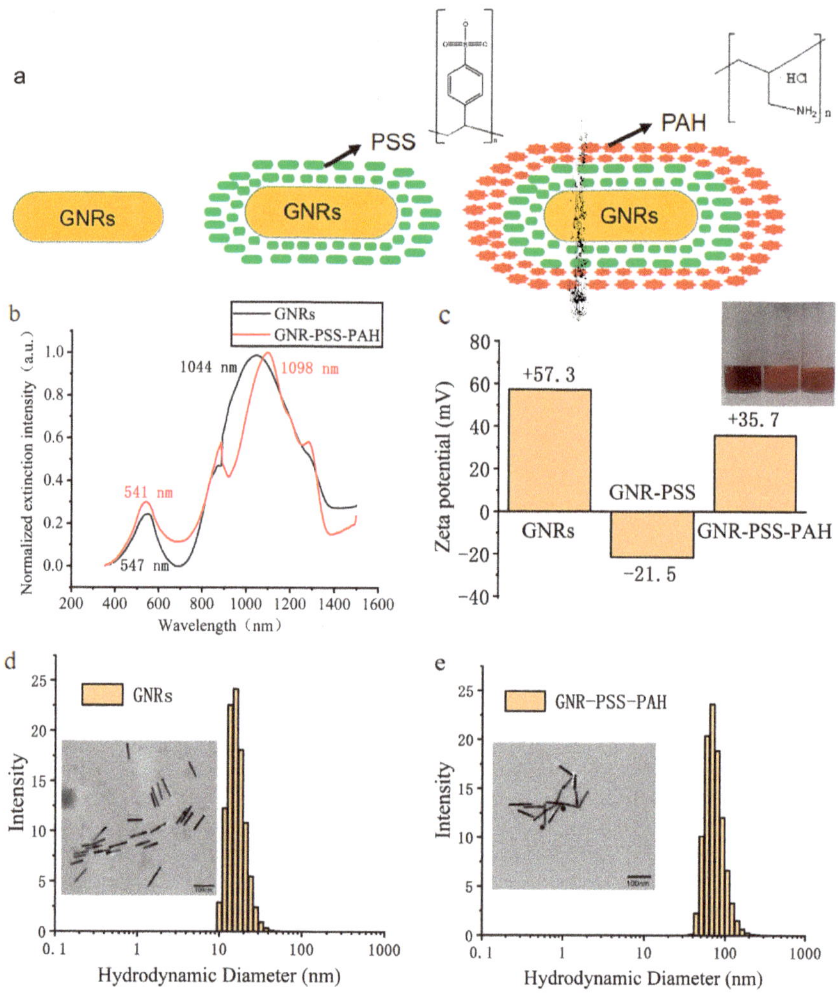

Figure 5. (**a**) Multilayer modification of gold nanorods with polymer electrolyte PSS and PAH. (**b**) Extinction spectra of GNRs and GNR-PSS-PAH. (**c**) Zeta potential of GNRs, GNR-PAA, and GNR-PSS-PAH. (**d**) DLS distribution and transmission electron microscope image (scale 100 nm) of GNRs. (**e**) DLS distribution and transmission electron microscope image (scale 100 nm) of GNR-PSS-PAH.

After incubating cells with GNR-PSS-PAH for 24 h, we acquired cell images at different wavelengths, as shown in Figure 6a–h. These images revealed that cell images obtained with GNR labeling were clearer. SBR analysis of the images (Figure 6i) indicated that the cell image SBR gradually increased with a redshift in the imaging wavelength for bands from 900 to 1700 nm, 1000 to 1700 nm, 1100 to 1700 nm, 1200 to 1700 nm, 1300 to 1700 nm, and 1400 to 1700 nm, with the highest SBR in the 1450 nm BP (1425–1475 nm) band, which is 32.4. The cross-sectional signal along the red dashed line of the cell in Figure 6d is presented in Figure S2, indicating a bright cytoplasm and relatively dark nucleus. Images from other fields are shown in Figure S3.

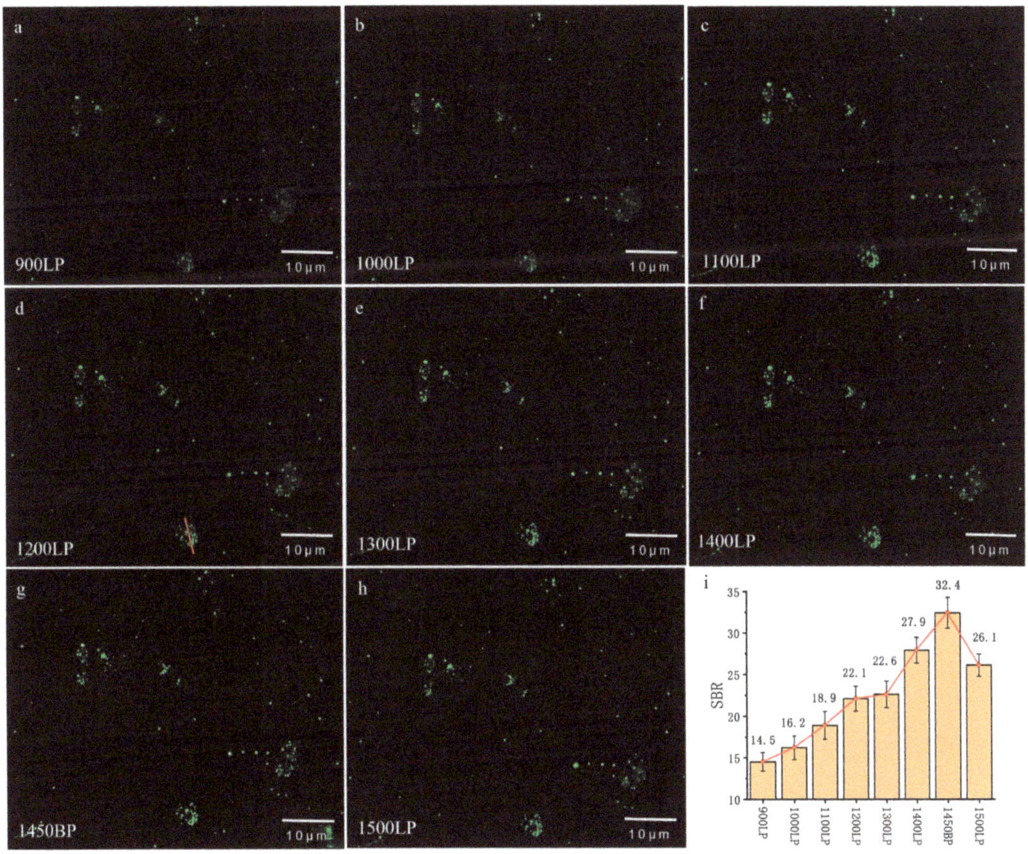

Figure 6. Dark-field imaging of cells in NIR-II window after incubation with gold nanorods. We added (**a**) 900LP, (**b**) 1000LP, (**c**) 1100LP, (**d**) 1200LP, (**e**) 1300LP, (**f**) 1400LP, (**g**) 1450BP, and (**h**) 1500LP to the system; (**i**) shows the comparison of SBR of the image above.

From the above results, it can be found that there was a four to six times improvement in SBR when using GNRs as contrast agents. The best images were observed in the 1450 nm BP (1425–1475 nm) band. This result is attributed to the reduced background noise due to water absorption near 1450 nm.

Moreover, dark regions were observed in the middles of the cells, indicating weaker scattering signals in the central region. This was because GNR-PSS-PAH could only enter the cell cytoplasm via endocytosis but not the cell nucleus. In the following experiments, we modified the gold nanorods with a nuclear localization sequence (NLS) to target the cell nucleus.

3.4. NIR-II Dark-Field Scattering Imaging of Nucleus-Specific GNRs

We used a nuclear localization sequence (NLS)-specific modification of gold nanorods to target the cell nucleus [30]. Figure 5a,b show the extinction spectra and zeta potential of GNRs and GNR-RGD-NLS, respectively, confirming the successful modification of RGD and NLS on the GNRs' surface. The LSPR extinction peak of GNRs shifted from 1106 nm to 1109 nm after modification (Figure 7a), with no significant change in the extinction spectrum shape, indicating successful RGD and NLS modifications. It can also be noticed that two peaks at about 900 nm and 300 nm were not supposed to appear, and they were actually due to the experimental setup and data analysis. The zeta potential measurements (Figure 7b) in water dispersion showed that GNRs had a zeta potential of +42.3 mV, which decreased to +4.8 mV after RGD and NLS attachment. The average hydrodynamic diameter of the GNRs was 17.4 ± 4.5 nm, and that of the GNR-RGD-NLS was 388.8 ± 77.3 nm, as shown in Figure 7c,d, respectively. The large difference was because the long chains of polymers would stretch out in water. The polymer dispersity index (PDI) of the GNRs and GNR-RGD-NLS was 0.07 and 0.04, which indicated a relatively uniform molecular weight distribution. These results further confirmed successful surface modifications.

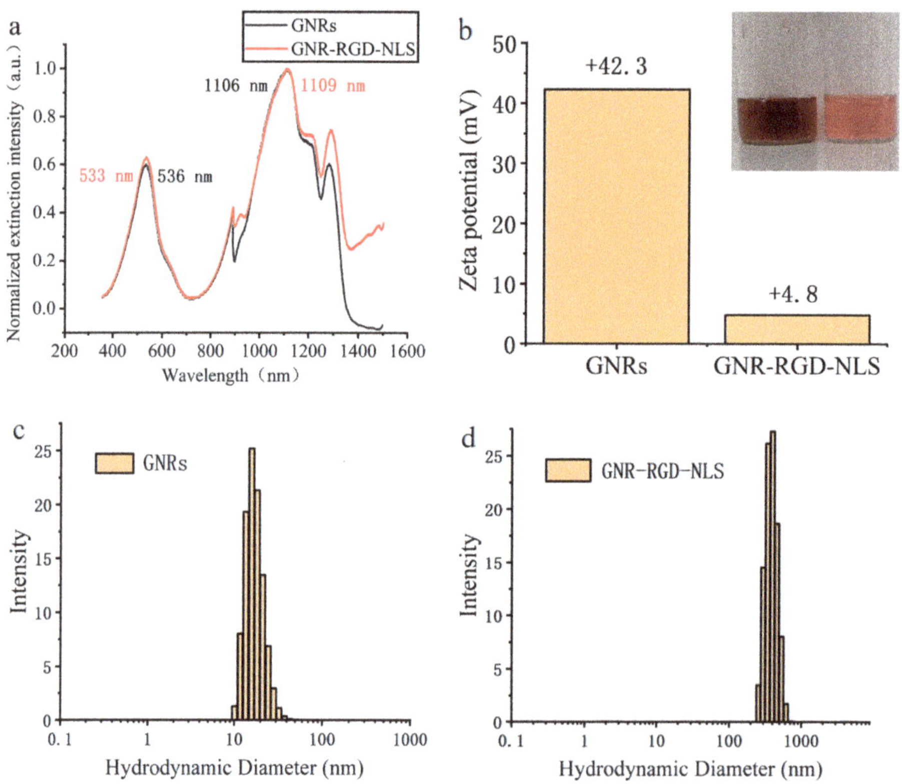

Figure 7. (**a**) Extinction spectra of GNRs and GNR-RGD-NLS. (**b**) Zeta potential of GNRs and GNR-RGD-NLS. (**c**) DLS distribution of GNRs. (**d**) DLS distribution of GNR-RGD-NLS.

Subsequently, dark-field microscopy enabled clear visualization of the cell nuclei labeled with GNR-RGD-NLS, exhibiting strong scattering signals in the bands of 1400–1700 nm (Figure 8c) and 1425–1475 nm (Figure 8d). The calculated image SBR values were 35.9 and 26.9, respectively. In addition, we stained the cell nuclei with propidium iodide (PI) to obtain fluorescence images (Figure 8a,b). The overlapped images of dark-field and fluorescence microscopy are shown in Figure 8e,f, with a false color of green for GNRs and red for PI. The overlap image of dark-field and fluorescence images indicated a high colocalization coincidence and the effective and specific labeling of the cell nuclei with GNR-RGD-NLS. Contrary to cells labeled with GNR-PSS-PAH in Figure S2, a bright nucleus was observed for cells labeled with GNR-RGD-NLS, and the cytoplasm was too dark to be seen, as shown in Figure 8g–j. This was because GNR-RGD-NLS could enter the cell nucleus due to the existence of the membrane-penetrating and nuclear localization proteins.

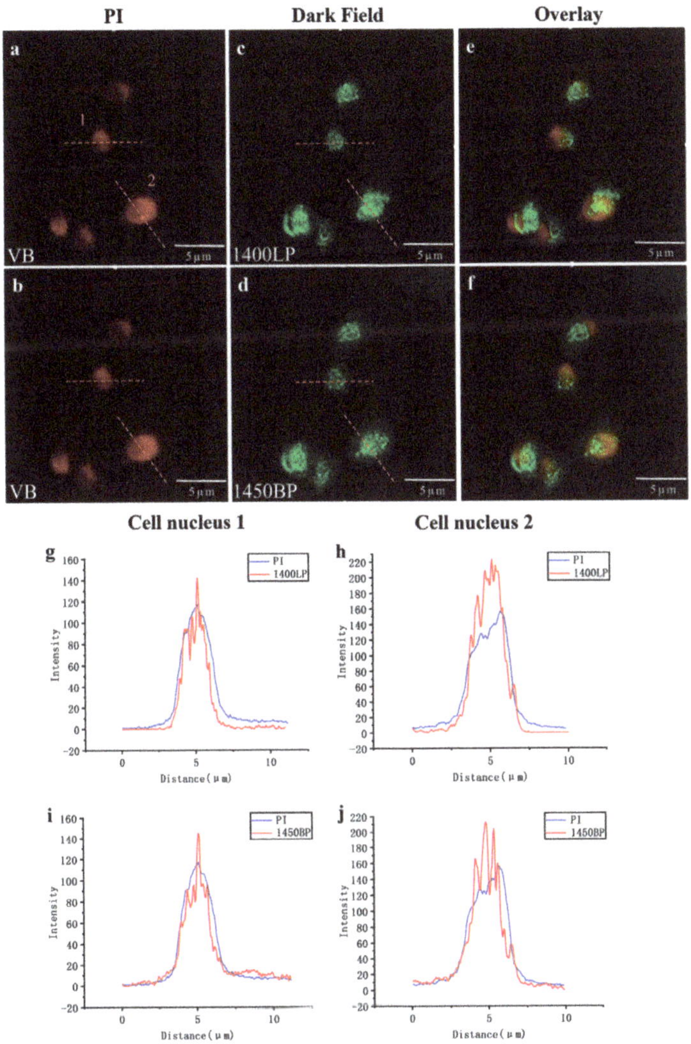

Figure 8. (**a**,**b**) The fluorescent image of cells stained with PI. (**c**) The dark-field image in the band of 1400–1700 nm of cells labeled with GNR-RGD-NLS. (**d**) The dark-field image in the band of 1425–1475 nm of cells labeled with GNR-RGD-NLS. (**e**) The overlapped image of (**a**,**c**). (**f**) The overlapped image of (**b**,**d**). The false color red was for PI and green was for GNRs. (**g**) The signal intensity distribution along the red dashed line of cell 1 in (**a**,**c**). (**h**) The signal intensity distribution along the red dashed line of cell 2 in (**a**,**c**). (**i**) The signal intensity distribution along the red dashed line of cell 1 in (**b**,**d**). (**j**) The signal intensity distribution along the red dashed line of cell 2 in (**b**,**d**).

Furthermore, we plotted the signal intensity distribution along the red dashed line of cell 1 in Figure 8a,c, as shown in Figure 8g. The same plotting method was conducted for cell 2, and the result is shown in Figure 8h. It was evident that the dark-field image of cells labeled with GNR-RGD-NLS exhibited stronger signals than that of PI-stained cells. The same plotting method was conducted for cells 1 and 2 in Figure 8b,d, as shown in Figure 8i,j. The result indicated that we also obtained high SBR cell images in the 1425–1475 nm wavelength range (1450 nm BP), commonly referred to as the NIR-IIx window [17]. Considering the significant water absorption and notable photothermal effects of this range, it is rarely used in biological fluorescence imaging. However, from another perspective, thanks to the significant absorption of water, the background noise from scattered photons was further suppressed. In our NIR-II dark-field Microscopy, there was no need to worry about the photothermal effect in this band due to the low photon density of the illuminating source, ensuring the safety of live cells.

4. Discussion

Researchers have defined the 900–1880 nm wavelength range as the NIR-II window. Imaging in this range has garnered significant interest in recent years due to its high image contrast and deep imaging capabilities. Dark-field microscopy, known for its affordability, non-destructive sample imaging, and minimal photobleaching, has found wide applications in cellular imaging in the visible band. This paper leverages the advantages of dark-field microscopy in conjunction with NIR-II imaging.

In this article, we primarily introduced a novel technical concept that extended the dark-field scattering microscopy into the near-infrared II region (900–1880 nm). Compared to the traditional visible dark-field scattering microscopy, the advantages of NIR-II dark-field scattering microscopy lie in suppressing the image background and enhancing the signal-to-background ratio, thereby obtaining clearer cell images. These advantages are effective for all types of cells, including but not limited to cancer cells. In this study, we confirmed this with a colon cancer cell line. Despite these advantages, there are still some disadvantages of NIR-II dark-field microscopy. One could be that it is not possible to obtain the bright field image of the same area at the same time as the contrast. In addition, the signal intensity of the samples decreases as the background intensity is suppressed. Therefore, we proposed the strategy of introducing gold nanorods as the scattering agents to enhance the signal intensity in the regions of interest. Compared to the traditional dark-field scattering imaging, our approach has a higher signal-to-background ratio and better specificity. By employing different modifications on the gold nanorods, we could image the cytoplasm and nucleus of cells separately, as we have demonstrated in this article. In future work, we can explore more modification methods for gold nanorods to obtain targeted images of specific organelles, thereby broadening the application fields of NIR-II dark-field microscopy.

5. Conclusions

We optimized a traditional dark-field imaging system to perform dark-field imaging of cells in the NIR-II range. Compared to images in the visible band, images in the NIR-II window exhibited reduced background noise, clearer cell contours, and increased SBR as the wavelength increased. To further increase the contrast of the cells, GNRs with an LSPR extinction peak in the NIR-II band were synthesized and modified with PSS and PAH. Dark-field images of GNR-PSS-PAH-labeled cells were most favorable in the 1400–1700 nm wavelength range with an SBR of 32.4, which was approximately six times higher than images without GNRs. In those images, the cytoplasm was bright while the nucleus was rather dark. Then, we modified GNRs with nuclear targeting proteins and obtained the nuclear-specific probe GNR-RGD-NLS. Images of GNR-RGD-NLS-labeled cells were obtained in both 1400–1700 nm and 1425–1475 nm bands, displaying well-defined nuclear boundaries. The colocalization coincidence with nuclear dye PI was very high, and the image in 1400–1700 nm had a high SBR of 35.9. These results underscored positive implications of our NIR-II dark-field microscopy in suppressing the image background and enhancing the signal-to-background ratio, thereby obtaining clearer cell images.

Supplementary Materials: The following supporting information can be downloaded at: https://www.mdpi.com/article/10.3390/biology12111391/s1, Figure S1: The cross-sectional signal along the red dashed line of the cell in Figure 4d, indicating a relatively uniform intensity of the cytoplasm and nucleus; Figure S2: The cross-sectional signal along the red dashed line of the cell in Figure 6d, indicating a bright cytoplasm and rather dark nucleus; Figure S3: Images from other fields. (a–c): Images of cells incubated with GNR-RGD-NLS.

Author Contributions: Conceptualization, Y.S., S.P. and W.L.; methodology, S.P., Z.H., Z.F., W.L., J.Q. and W.Z.; data curation, Y.S., S.P., Z.H. and W.L.; writing—original draft preparation, Y.S. and Z.H.; writing—review and editing, Y.S. and W.L.; funding acquisition, W.L. All authors have read and agreed to the published version of the manuscript.

Funding: This research was supported by Zhejiang Provincial Natural Science Foundation of China under Grant No. LQ21F050007, the National Natural Science Foundation of China (61975172 and 82001874), Fundamental Research Funds for the Central Universities (2020-KYY-511108-0007).

Institutional Review Board Statement: Not applicable.

Informed Consent Statement: Not applicable.

Data Availability Statement: The data that support the findings of this study are available from the corresponding author (Wen Liu) upon reasonable request.

Conflicts of Interest: The authors declare no conflict of interest.

References

1. Cao, J.; Zhu, B.L.; Zheng, K.F.; He, S.G.; Meng, L.; Song, J.B.; Yang, H.H. Recent Progress in NIR-II Contrast Agent for Biological Imaging. *Front. Bioeng. Biotechnol.* **2020**, *7*, 487. [CrossRef] [PubMed]
2. Kim, S.; Lim, Y.T.; Soltesz, E.G.; De Grand, A.M.; Lee, J.; Nakayama, A.; Parker, J.A.; Mihaljevic, T.; Laurence, R.G.; Dor, D.M.; et al. Near-infrared fluorescent type II quantum dots for sentinel lymph node mapping. *Nat. Biotechnol.* **2004**, *22*, 93–97. [CrossRef] [PubMed]
3. Ji, N. Adaptive optical fluorescence microscopy. *Nat. Methods* **2017**, *14*, 374–380. [CrossRef] [PubMed]
4. Horton, N.G.; Wang, K.; Kobat, D.; Clark, C.G.; Wise, F.W.; Schaffer, C.B.; Xu, C. In vivo three-photon microscopy of subcortical structures within an intact mouse brain. *Nat. Photonics* **2013**, *7*, 205–209. [CrossRef]
5. Domaille, D.W.; Que, E.L.; Chang, C.J. Synthetic fluorescent sensors for studying the cell biology of metals. *Nat. Chem. Biol.* **2008**, *4*, 168–175. [CrossRef]
6. Leung, C.W.T.; Hong, Y.N.; Chen, S.J.; Zhao, E.G.; Lam, J.W.Y.; Tang, B.Z. A Photostable AIE Luminogen for Specific Mitochondrial Imaging and Tracking. *J. Am. Chem. Soc.* **2013**, *135*, 62–65. [CrossRef]
7. Priest, L.; Peters, J.S.; Kukura, P. Scattering-based Light Microscopy: From Metal Nanoparticles to Single Proteins. *Chem. Rev.* **2021**, *121*, 11937–11970. [CrossRef]
8. Antolovic, V.; Marinovic, M.; Filic, V.; Weber, I. A simple optical configuration for cell tracking by dark-field microscopy. *J. Microbiol. Methods* **2014**, *104*, 9–11. [CrossRef]
9. Liu, H.; Dong, C.Q.; Ren, J.C. Tempo-Spatially Resolved Scattering Correlation Spectroscopy under Dark-Field Illumination and Its Application to Investigate Dynamic Behaviors of Gold Nanoparticles in Live Cells. *J. Am. Chem. Soc.* **2014**, *136*, 2775–2785. [CrossRef]
10. Li, H.B.; Wang, H.L.; Huang, D.S.; Liang, L.J.; Gu, Y.J.; Liang, C.Y.; Xu, S.P.; Xu, W.Q. Note: Raman microspectroscopy integrated with fluorescence and dark field imaging. *Rev. Sci. Instrum.* **2014**, *85*, 056109. [CrossRef]
11. Danilatos, G.; Kollia, M.; Dracopoulos, V. Transmission environmental scanning electron microscope with scintillation gaseous detection device. *Ultramicroscopy* **2015**, *150*, 44–53. [CrossRef] [PubMed]
12. La Spina, R.; Antonio, D.C.; Desmet, C.; Valsesia, A.; Bombera, R.; Norlen, H.; Lettieri, T.; Colpo, P. Dark Field Microscopy-Based Biosensors for the Detection of E. coli in Environmental Water Samples. *Sensors* **2019**, *19*, 4652. [CrossRef] [PubMed]
13. Qin, A.M.; Zhao, L.L.; Du, W.L.; Kong, D.X.; Qin, F.L.; Mo, R.W.; Zhang, K.Y. Luminescence Mechanism of Near-Infrared Quantum Dots. *Spectrosc. Spectr. Anal.* **2016**, *36*, 2059–2065.
14. Amiot, C.L.; Xu, S.P.; Liang, S.; Pan, L.Y.; Zhao, J.X.J. Near-infrared fluorescent materials for sensing of biological targets. *Sensors* **2008**, *8*, 3082–3105. [CrossRef] [PubMed]
15. Wang, Z.; Li, X.; Li, Y.Y.; Wang, Q.; Lu, X.M.; Fan, Q.L. Activatable NIR-II Probe for Tumor Imaging. *Prog. Chem.* **2022**, *34*, 198–206.
16. Hong, G.S.; Antaris, A.L.; Dai, H.J. Near-infrared fluorophores for biomedical imaging. *Nat. Biomed. Eng.* **2017**, *1*, 0010. [CrossRef]

17. Feng, Z.; Tang, T.; Wu, T.X.; Yu, X.M.; Zhang, Y.H.; Wang, M.; Zheng, J.Y.; Ying, Y.Y.; Chen, S.Y.; Zhou, J.; et al. Perfecting and extending the near-infrared imaging window. *Light Sci. Appl.* **2021**, *10*, 197. [CrossRef]
18. Welsher, K.; Sherlock, S.P.; Dai, H.J. Deep-tissue anatomical imaging of mice using carbon nanotube fluorophores in the second near-infrared window. *Proc. Natl. Acad. Sci. USA* **2011**, *108*, 8943–8948. [CrossRef]
19. Bruns, O.T.; Bischof, T.S.; Harris, D.K.; Franke, D.; Shi, Y.; Riedemann, L.; Bartelt, A.; Jaworski, F.B.; Carr, J.A.; Rowlands, C.J.; et al. Next-generation in vivo optical imaging with short-wave infrared quantum dots. *Nat. Biomed. Eng.* **2017**, *1*, 0056. [CrossRef]
20. Wang, R.; Li, X.M.; Zhou, L.; Zhang, F. Epitaxial Seeded Growth of Rare-Earth Nanocrystals with Efficient 800 nm Near-Infrared to 1525 nm Short-Wavelength Infrared Downconversion Photoluminescence for In Vivo Bioimaging. *Angew. Chem. Int. Ed.* **2014**, *53*, 12086–12090. [CrossRef]
21. Chang, K.W.; Sun, X.L.; Qi, Q.F.; Fu, M.Y.; Han, B.; Zhang, Y.; Zhao, W.; Ni, T.J.; Li, Q.; Yang, Z.J.; et al. NIR-II Absorbing Conjugated Polymer Nanotheranostics for Thermal Initiated NO Enhanced Photothermal Therapy. *Biosensors* **2023**, *13*, 642. [CrossRef]
22. Khan, N.U.; Lin, J.; Younas, M.R.; Liu, X.K.; Shen, L.M. Synthesis of gold nanorods and their performance in the field of cancer cell imaging and photothermal therapy. *Cancer Nanotechnol.* **2021**, *12*, 20. [CrossRef]
23. Nehl, C.L.; Hafner, J.H. Shape-dependent plasmon resonances of gold nanoparticles. *J. Mater. Chem.* **2008**, *18*, 2415–2419. [CrossRef]
24. Meng, L.Q.; Zhang, J.S.; Li, H.Q.; Zhao, W.W.; Zhao, T.G. Preparation and Progress in Application of Gold Nanorods. *J. Nanomater.* **2019**, *2019*, 4925702. [CrossRef]
25. Ke, S.L.; Kan, C.X.; Mo, B.; Cong, B.; Zhu, J.J. Research Progress on the Optical Properties of Gold Nanorods. *Acta Phys.-Chim. Sin.* **2012**, *28*, 1275–1290.
26. Link, S.; Mohamed, M.B.; El-Sayed, M.A. Simulation of the Optical Absorption Spectra of Gold Nanorods as a Function of Their Aspect Ratio and the Effect of the Medium Dielectric Constant. *J. Phys. Chem. B* **1999**, *103*, 3073–3077. [CrossRef]
27. Vigderman, L.; Zubarev, E.R. High-Yield Synthesis of Gold Nanorods with Longitudinal SPR Peak Greater than 1200 nm Using Hydroquinone as a Reducing Agent. *Chem. Mat.* **2013**, *25*, 1450–1457. [CrossRef]
28. Gole, A.; Murphy, C.J. Polyelectrolyte-Coated Gold Nanorods: Synthesis, Characterization and Immobilization. *Chem. Mat.* **2005**, *17*, 1325–1330. [CrossRef]
29. Pissuwan, D.; Niidome, T. Polyelectrolyte-coated gold nanorods and their biomedical applications. *Nanoscale* **2015**, *7*, 59–65. [CrossRef]
30. Pan, L.M.; Liu, J.A.; Shi, J.L. Nuclear-Targeting Gold Nanorods for Extremely Low NIR Activated Photothermal Therapy. *ACS Appl. Mater. Interfaces* **2017**, *9*, 15952–15961. [CrossRef]

biology

MDPI

Article

Characterization of Spontaneous Melanization by Fluorescence Spectroscopy: A Basis for Analytical Application to Biological Substrates

Anna Cleta Croce [1,2,*] and Francesca Scolari [1,2]

[1] Institute of Molecular Genetics, Italian National Research Council (IGM-CNR), Via Abbiategrasso 207, I-27100 Pavia, Italy
[2] Department of Biology & Biotechnology, University of Pavia, Via Ferrata 9, I-27100 Pavia, Italy
* Correspondence: croce@igm.cnr.it; Tel.: +39-0382-986-428

Simple Summary: Melanin is a pigment naturally present in various living beings, from animals to plants and fungi. Melanin can be involved in many processes, such as innate immunity and antioxidant protection, including photoprotection and the related abilities to confer color to many biological structures and to interact with light. Light interaction, in particular, is largely exploited to investigate melanin chemical components and precursors for various purposes, from the in situ grading of the pigment for diagnosis in biomedicine to the management of light energy for the development of innovative applications in bioengineering. In this study, the ability of melanin and its precursors to respond to light irradiation have been characterized in terms of changes in the emission light signals, namely fluorescence. Besides the confirmation of the fluorescence emission signature of melanin and of its precursor L-tyrosine, new insights have been provided on intermediates during melanin production and by different conditions of measurement. The subsequent characterization of aqueous extracts from black cat hairs and mosquito eggs proved the value of these experimental procedures for application in the investigation of melanization processes in biological substrates.

Abstract: Melanin is present in various biological substrates where it may participate in several processes, from innate immunity to the still-unsolved opposite roles in antioxidant protection, including photoprotection and the related ability to interact with light. Melanin–light interaction has also been an important source of inspiration for the development of innovative bioengineering applications. These are based on melanin's light-energy-absorption ability of its chemically and structurally complex components and precursors, and on the improvement in analytical and diagnostic procedures in biomedicine. In this regard, here, we characterized the fluorescence spectral properties of melanin and of its precursor L-tyrosine in an aqueous solution during spontaneous melanization. Besides the confirmation of the typical fluorescence-emission signature of melanin and L-tyrosine, we provide additional insights on both emission and excitation spectra recorded during melanization. On these bases, we performed a subsequent characterization on the aqueous extracts from two different melanin-containing biological substrates, namely hairs from a domestic black cat and eggs from the Asian tiger mosquito. The results from the mild extraction procedure, purposely applied to obtain only the soluble components, combined with fluorescence spectral analysis are expected to promote further investigation of the melanization processes, particularly in insects.

Keywords: emission spectra; excitation spectra; melanin; L-tyrosine; Asian tiger mosquito; *Aedes albopictus* eggs; black cat hairs

1. Introduction

Melanin is a natural biopolymer ubiquitously distributed in most living beings, from fungi to different taxa of plants and animals [1–3]. Abundant amounts of melanin are usually obtained from the common cuttlefish, *Sepia officinalis,* to be used for multiple purposes, from the study of its composition and biophysical properties to pharmaceutical and industrial applications or for alimentary use [4,5]. The generally known ability of many cephalopods to produce and eject melanin in their natural environment as an ink for defense and interaction between prey and predator is a primary example of the various biological functions played by melanin and its precursors, also considering that melanin shares similar intermediate products with biomolecules acting as hormone regulators [6]. The photoprotection and tanning of human skin are within the generally recognized effects of melanin [7]. In addition, hypotheses about the contrasting roles of melanization processes in the development of skin cancer and melanomas, as well as in the protection from such diseases, have been formulated [6,8,9]. In higher animals, neuromelanin produced from the oxidation of dopamine and other catecholamines in dopaminergic neurons accumulates in defined sites of the central nervous system, such as in the *substantia nigra* and in the *locus coeruleus,* where positive or negative outcomes may depend respectively on its ability to bind metal ions and toxic compounds or to leak from dying neurons and activate microglia, inflammation and neurodegeneration [10,11]. Melanin can also act as antimicrobial and antiparasitic agent, contributing to the innate immune response in the skin and various tissues of humans, as well as in different animals including insects [12–16].

The presence of melanin and its precursors in different body sites in an organism is thus relevant for its various, multifaceted, interlaced and still not completely characterized biological roles, which is, in turn, relatable to its biochemical, photophysical and antioxidant properties [1–3,17,18].

The production of melanin from tyrosine occurs through oxidative reactions catalyzed by the enzyme tyrosinase or spontaneously in the presence of oxygen, with the subsequent aggregation of oligomeric intermediates. Among the possible various intermediates, 5,6-dihydroxyindole and 5,6-dihydroxyindole-2-carboxylic acids have been identified as the precursors able to combine in macromolecules with a complex and disordered secondary structure in dark brown eumelanin. The involvement of cysteine for the synthesis of cysteinyl-DOPA is distinctive in the production of the reddish pheomelanin, less photo-stable and more pro-oxidant than eumelanin [6,18,19].

The melanization process entails the production of many chemically heterogenous intermediates. Analytical procedures like HPLC, electrospray ionization mass-spectrometry, matrix-assisted laser desorption/ionization mass-spectrometry or pyrolysis gas chromatography, as well as ultrastructural imaging techniques, have been applied to detect specific oxidative or hydrolytic degradation products of melanin to study synthetic melanogenesis and its regulation in cells, or to investigate the structural organization of natural melanin [19–26]. The comparison of data from various studies that has been proposed is hardly able to provide equivalent analytical results, which can provide complementary information useful to characterize melanin of various biological substrates for different applications [26]. In any case, the above-mentioned analytical procedures are complex, costly, time-consuming and limited to the defined time points of the sample collection. Conversely, optical spectroscopy and imaging, although unable to specifically detect single intermediate species, are suitable to perform real-time, in situ, repeatable assays to monitor melanization and detect melanin.

The well-assessed melanin light-absorption properties consist in a continuous curve which decreases monotonically from near-UV along the visible spectrum. As for fluorescence, the increase in the excitation wavelengths results in melanin emission curves sharing their boundary at their longest wavelength side, with peaks localized at progressively longer wavelengths and with lower amplitudes [27–29].

Contrarily to melanin, its chemical precursor L-tyrosine exhibits well-defined and narrower absorption and fluorescence excitation and emission bands. The optical properties of L-tyrosine have been initially characterized along with those of phenylalanine and tryptophan in the near-UV–blue spectral region, to provide a basis for further studies on peptides and proteins [30–32]. These three aromatic amino acids share similar absorption/excitation optical profiles with two bands, the major one peaking around 220 nm, and a minor and broader one peaking at about 260 nm, 275 nm and 280 nm for phenylalanine, L-tyrosine and L-tryptophan, respectively. Emission spectra, in turn, respectively peak around 280 nm, 300 nm and 350 nm, indicating that the different substituents of these aromatic amino acids affect the absorption/excitation spectral profiles less than the fluorescence emission properties [30]. A red emission band above 600 nm for both L-tyrosine and L-tryptophan has been also reported [33].

Up to now, absorption spectroscopy has been the foremost optical technique used to follow the production and aggregation of intermediates during the melanization process starting from the oxidation of tyrosine and dopamine, as well as the presence of specific components. Different experimental and computing studies based on spectroscopic simulation have shown the progressive increase in and broadening of the absorption bands in the 270–350 nm and 400–600 nm spectral ranges [33–38]. Fluorescence investigation has considered only some specific indole derivatives as precursors or components of eumelanin, with their possibly related emission and excitation bands [29,39,40].

Different kinds of diagnostic and analytical procedures are increasingly considering the optical properties of melanin as a useful intrinsic biomarker. In histopathology, melanin can hinder the bright-field-microscope observation of morphology and expression markers of stained and immune-stained cells and tissues. This problem can be solved with a suitable light irradiation of the tissue slice to induce photobleaching and the consequent disappearance of the dark pigment, as verified for melanoma samples from different veterinary sources [41]. Conversely, sophisticated techniques such as multiphoton excitation and lifetime spectroscopy and imaging focus on melanin as an optical parameter for the real-time, in vivo screening of skin disease and diagnosis of melanoma [42–47]. Similarly, in ophthalmology, melanin is characterized along with lipofuscins for the in vivo optical detection and monitoring of retinal aging and diseased conditions [48–50].

In insects, the optical properties of melanin are widely considered for studying the melanization processes, considering its importance in immunity, as well as the role of pigment contribution to the body biocomponents relevant to the hardening of the superficial cuticle and pigmentation patterns, with related behavioral implications [3,14,51–54]. In a recent study, melanins have been suggested to contribute to the autofluorescence of the reddish components of some brownish scales covering the body of the Asian tiger mosquito, *Aedes albopictus* (Skuse, 1894) (Diptera, Culicidae) [55]. These findings opened novel opportunities to the investigation of mosquito optical properties, a recently expanding research field [55–58], either to develop new tools for vector tracing or to shed light on potential fluorescence-based communication strategies in mosquitoes.

In this context, the aim of this work is to set up a simplified, optically based procedure to help the investigation of the melanization processes in different biological substrates. To this end, spectroscopy was applied to assess fluorescence excitation and emission changes in L-tyrosine undergoing spontaneous oxidation in an aqueous solution, up to the formation of synthetic eumelanin. The experimental excitation and emission spectra of L-tyrosine, L-tryptophan and melanin were recorded and are shown in combination as reference data to allow for an easier and more direct appreciation of the spectral shape changes occurring during melanization.

Throughout this study, eumelanin will be named "melanin". L-tryptophan was also considered, because its presence in biological substrates may contribute to their fluorescence. The spectroscopic results from these synthetic compounds were then used for a comparative investigation on the presence of melanin intermediates in aqueous solutions obtained by a mild extraction procedure from the hairs of a black domestic cat and the eggs of *Ae. albopictus*, as natural sources of melanin in phylogenetically distant animal species [59–61].

2. Materials and Methods

2.1. Chemicals

L-tyrosine and L-tryptophan were obtained as dry powders (both BioUltra-≥99.0%-degree, Sigma Chem Co. St Louis, MO, USA) to be dissolved in phosphate buffer (pH 7.2, 0.1 M) as saturated stock solutions (5.0 mg/mL and 11 mg/mL, respectively). The solutions were preserved in the dark, in humidified air to allow exposure to atmospheric oxygen and consequent oxidation of the amino acids. A 24-month-old solution provided synthetic melanin, as a completed melanization product.

2.2. Biological Samples

Black hairs were collected by brushing a domestic black cat (*Felis catus* Linnaeus, 1758) with a soft brush. Mosquito eggs from the *Ae. albopictus* RER strain [62], which was obtained from the Centro Agricoltura Ambiente (CAA) "G. Nicoli" (Crevalcore, Italy) in 2022, were collected (about n = 30 eggs) and immediately processed for the aqueous extract preparation. The biological samples were rinsed three times with ultrapure water before the further processing with the gentle mild procedures, purposely applied to preserve the chemical and structural properties of the investigated biomolecules.

Fresh biological samples were transferred to an isotonic solution (NaCl 0.9%; 2 mL) to be homogenized using an immersion homogenizer (Ultra-Turrax T25, IKA Werker, Janke & Kunkel, Breisgau, Germany). The solutions were centrifuged, and the supernatant was collected to be preserved at −20 °C until they were subjected to spectrofluorometric analysis. For each sample, the procedure lasted a total of two minutes. At the end of each preparation, the homogenizer-dispersing tool was removed and cleaned following three cycles of sonication (5 min each) and rinsing with ultrapure water.

Pictures of both mosquito eggs and black cat hairs were collected using a stereomicroscope (Olympus SZ40, Olympus Optical Co. GmbH, Hamburg, Germany), coupled with a smartphone camera.

2.3. Spectrofluorometric Analysis

The fluorescence excitation and emission spectra were recorded from the L-tyrosine and L-tryptophan solutions immediately after their preparation and at different times from the L-tyrosine solution preparation, or from the biological sample solutions by means of the LS 55 Model spectrofluorometer (PerkinElmer Italia, Milan, Italy). All measurements were performed at room temperature. The solutions from the pure compounds were subjected to consecutive sets of measurements by varying the excitation and emission conditions, to detect the different excitation and emission bands and define the related measurement conditions to be applied for the subsequent spectrofluorometric study of the biological sample preparations. To minimize inner-filter effects, the optical density at the spectral positions of interest was checked by a spectrophotometer (Spectronic Genesys Spectrophotometer, Spectronic Instruments Inc., Rochester, NY, USA), and proper dilutions were prepared at the convenience of the linearity of the fluorescence response.

The emission bands detected from the biological sample solutions by exciting at 220 nm were analyzed by means of a curve-fitting procedure, to validate and estimate the effective contribution of L-tyrosine and L-tryptophan. The Marquardt–Levenberg algorithm [63] was applied using the PeakFit program (Peak fit; SPSS Science, Chicago, IL, USA) first to define the parameters representing the emission bands of L-tyrosine and L-tryptophan in terms of respective Half-Gaussian Modified Gaussian (GMG) spectral functions. The GMG functions were then used to analyze the emission of the biological samples, to detect and estimate the relative contribution of each amino-acid emission to the overall band area. More details on the procedure are provided in the Supplementary Materials.

3. Results

Synthetic Melanin and Pure Compounds

Synthetic melanin in solution, excited at wavelengths ranging from 230 to 500 nm, shows a series of emission curves. The increase in the excitation wavelength results in a shift in the peak positions from about 440 nm to 560 nm, and in a parallel decrease in the signal amplitudes.

Only the shortest excitations at 230 nm and 270 nm result in the presence of a minor band centered at about 360 nm (Figure 1A). Excitation profiles consist in a band centered at about 280 nm, when recorded at 320 nm and 360 nm. Observations at longer wavelengths show signal profiles decreasing from 200 nm to 290 nm, as well as signal profiles covering increasingly longer spectral ranges depending on the lengthening of the recording wavelength.

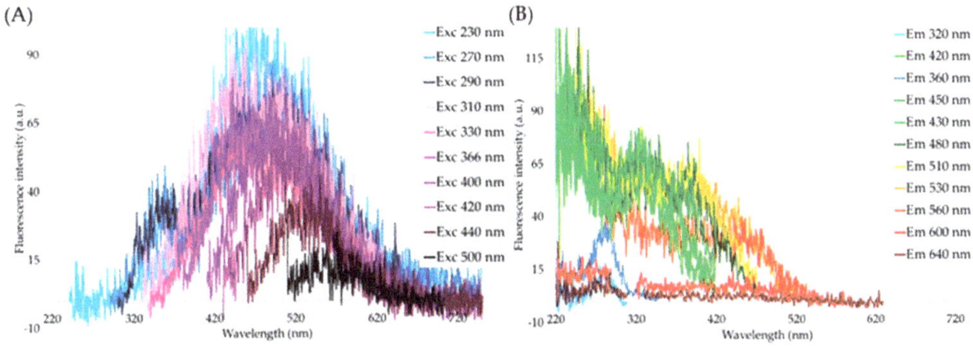

Figure 1. Emission (**A**) and excitation (**B**) spectra recorded from synthetic melanin in solution. Excitation and emission conditions are indicated by colors, on the right.

A band covering the range between 295 nm and 400–420 nm is observed at both 420 nm and 430 nm. An additional band in the 370–500 nm range is observed at 510 nm and 530 nm, and a lower and less structured emission profile is recorded at 560 nm. Observations at longer wavelengths, from 600 nm to 640 nm, show a very low signal profile in the 220–300 nm range, followed by an almost negligible emission at longer wavelengths (Figure 1B).

Following the settlement of the optical features of synthetic melanin in solution, the fluorescence excitation and emission properties of L-tyrosine were recorded to provide initial reference measurement conditions for the subsequent study of optical changes occurring during melanization (Figure 2A).

Emissions excited at both 230 nm and 270 nm consist in a band in the 280–350 nm range, peaking at about 310 nm. An additional band is detected in the red region, with the maximum around 610 nm. Excitation, in turn, consists in a signal decreasing rapidly from 220 nm to a minimum at 240 nm, followed by a band centered at about 270 nm. This profile is observed at both 310 nm and 610 nm, while negligible fluorescence signals are obtained under different excitation and emission conditions (Figure 2A). In comparison with the spectra of L-tyrosine, L-tryptophan shows similar fluorescence-excitation profiles, while emission spectra recorded in both the blue and red regions are wider and red-shifted, with the respective maximum values at about 360 nm and 860 nm (Figure 2B).

The fluorescence spectra recorded from L-tyrosine solution at one month of aging show the typical bands observed from the freshly prepared solution, accompanied by some changes in their profile and by the appearance of new spectral profiles. In more detail, excitation at 220 nm and 270 nm still results in the typical and well-defined emission bands centered at about 310 nm. Excitations longer than 300 nm result in a series of emission curves in the 350–600 nm spectral range, which in general undergo a decrease in emission amplitude depending on the lengthening of the excitation wavelength. In addition, excitation from 310 nm to 340 nm results in emission bands peaking at about 420 nm, while excitation at longer wavelengths results in an emission profile widening up to about 590 nm (Figure 3A).

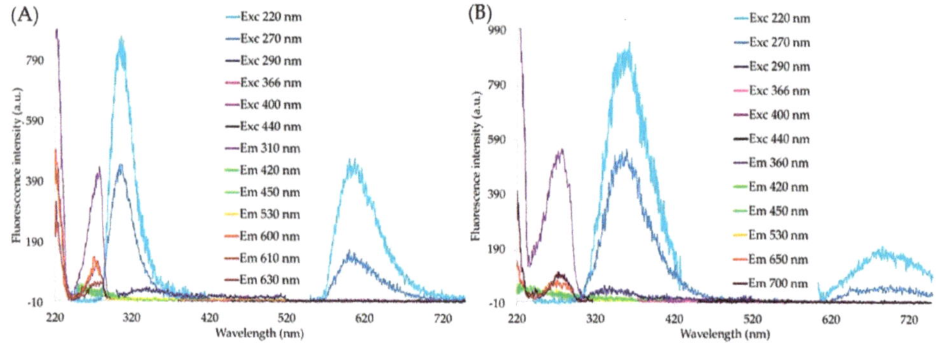

Figure 2. Emission and excitation spectra recorded from L-tyrosine (**A**) and L-tryptophan (**B**) freshly prepared solutions. Excitation and emission conditions are indicated by colors in the inlet legend.

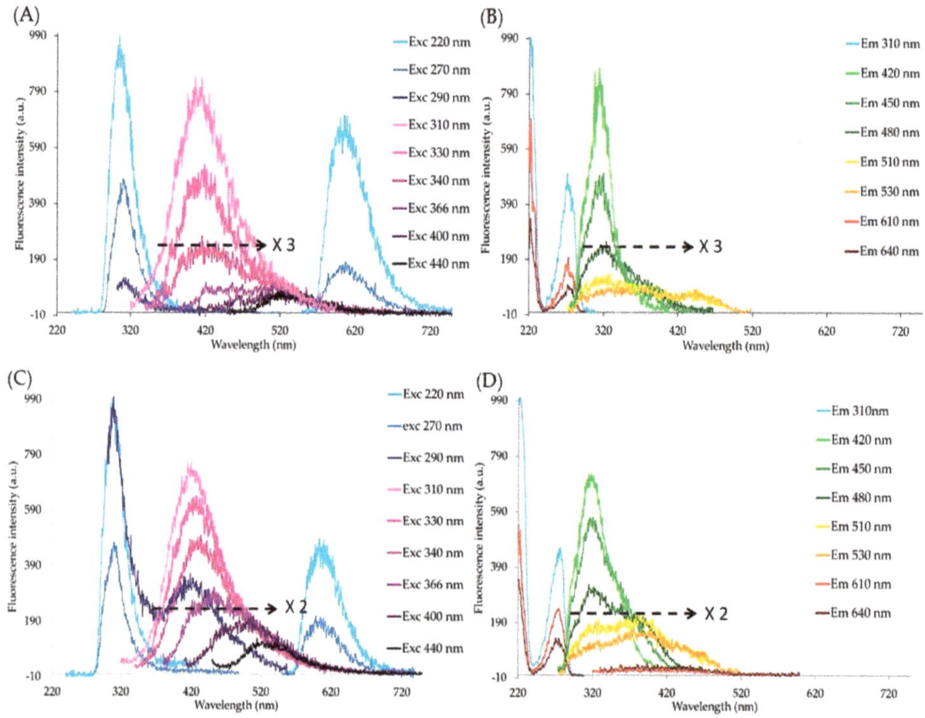

Figure 3. Emission (**A**,**C**) and excitation (**B**,**D**) spectra recorded during the melanization processes, at one (**A**,**B**) and six months (**C**,**D**) of L-tyrosine solution aging. To allow the shapes of the spectra covering the 320–620 nm range (**A**) and the 300–520 nm range (**B**) to be better appreciated, their amplitudes are increased with respect to the fluorescence intensity scale values, as indicated by the arrow in each panel. Excitation and emission conditions are indicated by colors on the right.

Excitation profiles recorded at 310 nm, 610 nm and 640 nm are comparable to those detected from L-tyrosine, while additional bands appear in the 320–520 nm range when the signal is detected at longer wavelengths. In particular, a well-defined band peaking at about 320 nm is observed at 420 nm. This band shows a decrease in its amplitude when the recording wavelength is lengthened up to 530 nm, as well as the appearance of shoulders centered at about 380 nm and 445 nm (Figure 3B).

The fluorescence spectra recorded from the L-tyrosine solution at six months of aging still show the typical bands of L-tyrosine, in the UV- and red-emission regions. The curves detected in the 350–600 nm spectral range under increasing excitation wavelengths show a constant red-shift in their maximum peaks, in parallel with a decrease in their amplitude. Notably, these emission bands share a common boundary position at the longest wavelength, similarly to what is observed for the synthetic melanin (Figure 3C). Excitation spectral profiles are similar to those described for the one-month-aged L-tyrosine, except for the increase in the amplitude of the bands peaking at 380 nm and 445 nm when observed in the 420–530 nm range (Figure 3D).

Fluorescence spectroscopy revealed a series of emission and excitation bands with different peak positions and spectral profiles, in a close dependence on the changes in the respective excitation and observation wavelengths. A sensitive and more variegated response is thus provided by fluorescence analysis, compared with the smooth spectral profiles of the absorption spectra, as shown by our direct measurements (Supplementary Materials, Figure S1) and by the literature [28,38].

The spectroscopic analysis of the pure compound solutions was followed by the characterization of the aqueous extract preparations obtained from the two different biological samples, namely the black cat hairs (Figure 4A) and the eggs of *Ae. albopictus* (Figure 4B).

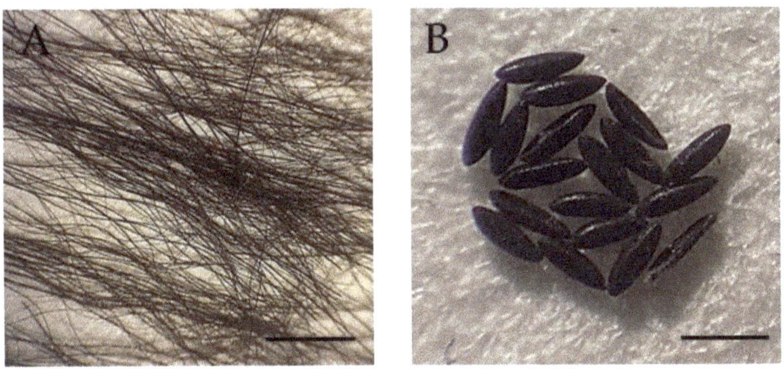

Figure 4. Images of (**A**) black cat hairs and (**B**) mosquito eggs. Bars: 600 µm (**A**), 500 µm (**B**).

Fluorescence spectra recorded from the solution prepared from the black cat hairs show three well-defined emission bands peaking at about 340 nm, and two bands in the 600–720 nm interval, when excited in the 220–290 nm range. Much lower emission profiles, with maxima at about 420 nm, are observed under excitation at wavelengths longer than 310 nm (Figure 5A).

Excitation spectra observed at both 310 nm and 340 nm or in the red region show a signal decreasing sharply from 220 nm to the minimum at about 240 nm, and a well-defined band peaking at about 270 nm. The excitation profiles observed in the 420–530 nm range show maximum peaks at about 320 nm, followed by a slow signal decrease up to 420 nm (Figure 5B).

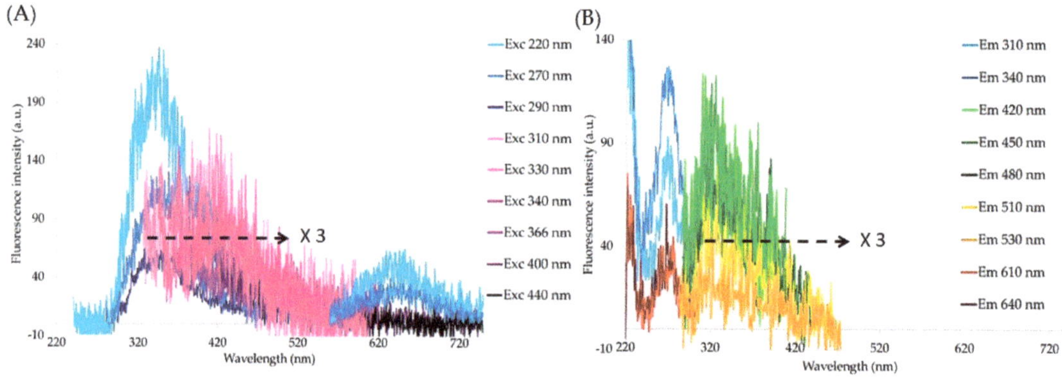

Figure 5. Emission (**A**) and excitation (**B**) spectra recorded from the preparation extract from black cat hairs. To allow the shapes of the spectra covering the 320–620 nm range (**A**) and the 300–470 nm range (**B**) to be better appreciated, their amplitudes are increased with respect to the fluorescence intensity scale a.u. values as indicated by the arrow in each panel. Excitation and emission conditions are indicated by colors on the right.

The aqueous extract solution obtained from the mosquito eggs shows spectra similar to those measured from black cat hairs in the shorter spectral region, or in the red one. Much more defined excitation and emission profiles are detected in the 320–420 nm or in the 400–600 nm range, respectively. Interestingly, excitation at wavelengths longer than 310 nm results in a series of emission spectra showing a red shift in the maximum peak accompanied by a decrease in the amplitude, similarly to what is observed for both synthetic melanin and for the advanced melanization step (Figure 6A).

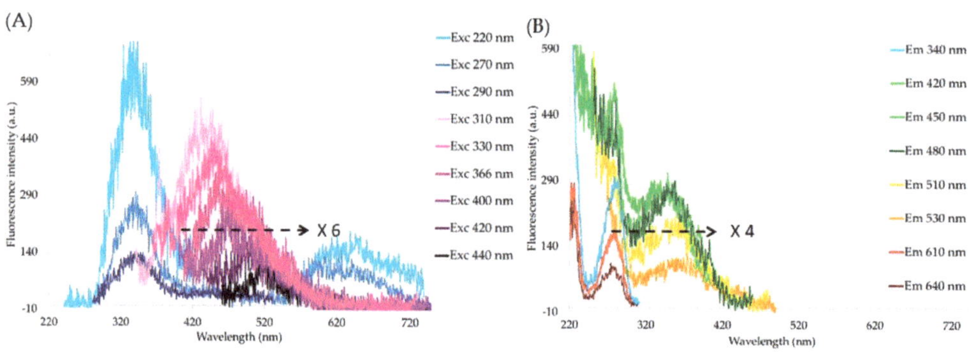

Figure 6. Emission (**A**) and excitation (**B**) spectra recorded from the preparation extract from mosquito eggs. To allow the spectral shapes of the spectra covering the 320–620 nm range (**A**) and the 300–520 nm range (**B**) to be better appreciated, their amplitudes are increased with respect to the fluorescence intensity scale a.u. values as indicated by the arrow in each panel. Excitation and emission conditions are indicated by colors on the right.

Excitation spectra exhibit a shift in the maximum values from about 340 nm to 360 nm depending on the lengthening of the observation, i.e., from 420 to 530 nm (Figure 6B).

Lastly, attention was given to the main emission bands of the biological samples excited at 220 nm, which are slightly red-shifted and wider compared to the emission of L-tyrosine, consistent with a possible contribution of L-tryptophan. These emission bands were thus subjected to a spectral fitting analysis procedure, which confirmed that both amino acids participate in the emission spectra excited at 220 nm (Supplementary Materials, Figures S2 and S3). The relative areas covered by the signals ascribed to the two amino acids were also estimated, indicating that L-tryptophan contributes to the overall signal for a fraction slightly greater than 50% in both biological samples, while the relative contribution of L-tyrosine is about 28% in mosquito eggs and 13% in black cat hair solution extracts.

4. Discussion

The fluorescence emission spectra collected from L-tyrosine in freshly prepared solutions are fully consistent with their specific optical signatures described in the literature [30,31,33,64]. Additionally, synthetic melanin shows its typical continuous series of fluorescence-emission curves, characterized by the peaks undergoing a red shift and the amplitude lowering at increasing excitation wavelengths while sharing the same boundary at their red side. This is in agreement with the literature results on melanin fluorescence excited at wavelengths from 360 nm to 400 nm [27–29]. These optical results are commonly considered to be in keeping with the exponential decrease in the melanin-absorption-curve profile, interpreted in terms of a linear combination of successive, partially superimposed and individually unresolvable Gaussian curves. These curves, in turn, have been proposed to reflect comparable, although chemically different, species, with different polymerization degrees accounting for a progressive weakening of the strength of energy transitions at increasing measurement wavelengths [65]. In this respect, it is also worth recalling that, similarly to humic acids, the continuous and broad absorption spectrum of melanin could result from a system of continuous electronic levels, favorable for electronic excitation and transitions relevant to the induction of the "chemiluminescence" events activated by the photo-oxidation of synthetic melanin [66]. Subsequent works have substantiated the concept of the common system of electronic levels resulting from the disordered network of aromatic species, able to favor the energy transitions matching with the absorption and emission of light in the visible range [40,65]. A more recent work based on ultrafast laser spectroscopy on the induction of excitons and charge transfer in the heterogenous chromophores of melanin have even compared the energetic behavior of melanin to that of a disordered graphene-like material [67]. In this respect, it is also worth recalling that the manipulation of the self-assembly process with changes in ions and pH in the production of liquid-like melanin have been explored, posing promising perspectives for innovative applications in bioengineering [68]. All the above recalled studies thus substantiate the extreme complexity of melanin, which does not affect its typical, well-assessed spectral signatures.

As for the excitation spectra, our synthetic melanin showed a sequence of excitation spectra which widen at their red side, depending on the lengthening of the observation wavelength. This finding is consistent with the few reports provided in the literature on melanin excitation spectroscopy [45]. In addition, the changes in the excitation profile, with the signal decrease from 220 nm and the subsequent appearance of two bands at longer spectral positions when lengthening the observation wavelength, recall the behavior shown by Teuchner and colleagues, who also verified that melanin fluorescence depended on solvent medium and on single- or two-photon excitation [45].

The initial oxidation of L-tyrosine and the subsequent production and aggregation of intermediates during the melanization process result in progressive changes in its optical properties. Different absorption-based studies have provided a general indication on the increase and widening of the bands in the 270–350 nm and 400–600 nm spectral ranges detected after the irradiation or oxidation of L-tyrosine and DOPA derivatives [33,36–38]. Comparable data have been computed by means of spectroscopic simulation based on the structure and energy status of species such as indole quinone reduced forms and their oligomers [34,35,69]. As for fluorescence, information is limited to few, specific indole-based intermediates. For example, the steady-state fluorescence emission spectrum of 5,6-dihydroxyindole has been found to exhibit a main band peaking at about 340 nm, as well as a band in the 500–600 nm spectral range detected from time-resolved spectra [39]. A synthetic N-methyl,5-hydroxy,6-methoxyindole homopolymer exhibited two wide-emission bands, one peaking at a wavelength slightly shorter than 500 nm when excited at 355 nm, and one peaking at slightly over 500 nm, independently from the increase in the excitation wavelength from 370 to 400 nm [40]. The collection of excitation spectra with a fixed peak at around 365 nm independent from the increase in the reading wavelengths suggests the presence of dihydroxyindole-carboxylic acid in the monomeric state or slightly bound to a synthetic eumelanin polymer [29].

Our data collected at intermediate stages of melanization have shown the appearance of new, well-defined emission bands in the 350–600 nm spectral range, rather than a widening of the still-present bands typical of L-tyrosine. In particular, the 366 nm excitation results in a wide emission covering the 420–520 nm range, while both 400 nm and 440 nm excitations result in two bands peaking at about 520 nm, similarly to what is reported in the literature for the N-Me-5H6MI homopolymer [40]. Excitation spectra, in turn, cover a 290–520 nm range, nearly comparable to what is reported in the literature for eumelanin [29]. In our case, anyway, observation at 420 nm reveals a 320 nm excitation band. The amplitude of this band decreases upon the lengthening of the measuring wavelength, accompanied by the rising of a shoulder at about 380 nm and of secondary band centered at about 440 nm. The changes in both excitation and emission profiles become more evident at longer time periods, after the preparation of the L-tyrosine solution, making them more similar to what is detected from the synthetic melanin solution. The typical bands of L-tyrosine, however, are hardly detectable in the synthetic melanin solution, indicating its almost complete transformation.

The above-described fluorescence data from the pure compound preparations, along with the available literature on the presence of melanin in biological samples [60,70], have been used to interpret the spectra from the aqueous extract solutions prepared from mosquito eggs and black cat hairs. The solution obtained from the mosquito eggs excited at wavelengths ranging from 310 to 440 nm showed well-defined emission spectra in the 350–600 nm range, in parallel with excitation bands in the 320–420 nm interval observed at wavelengths varying from 420 nm to 530 nm. All these results are in keeping with the presence of melanin compounds in the mosquito eggs, as previously shown in *Ae. albopictus* [70]. The preparation from black cat hairs, in turn, shows much lower signals in the spectral region typical of melanin compounds. The maxima of all emission spectra occur at about 420 nm, and the maxima of excitation spectra at about 320 nm. Excitation spectra also show a rapid decrease in the signal at their red side, independently from the lengthening of the observation wavelength, and can be easily distinguished from those relevant to L-tyrosine.

The comparison of the fluorescence results from the extract solutions prepared from the two different biological samples has revealed a remarkable difference as to the spectral region relating to melanin components. In this respect, it worth recalling that all our measurements were performed on aqueous solution extracts, similarly to the pure-compound preparations. The relatively gentle method adopted for the sample extraction was purposely chosen to obtain the most easily solubilizing compounds, preserving their chemical properties as much as possible compared to more aggressive treatments used for melanin isolation. In fact, these harsh methods can involve severe chemical treatments, such as extensive hydrolysis with boiling mineral acids or bases and followed by organic solvents' washing to isolate melanin formed inside melanosomes, which are bound to cellular components such as minerals and proteins, and may result in alterations in the chemical and structural properties of melanin [24,26,61]. On these bases, the melanin tightly bound to the proteinaceous components of the hair's mature melanosomes can likely hinder its extraction by our mild isolation method, thus accounting for our autofluorescence results on melanin extracts from black cat hairs [61,71].

Apart from the data of the spectral region relevant to melanin, both biological samples show the main emission band recorded by exciting at 220 nm, which is slightly red-shifted and wider than the emission of L-tyrosine. This finding indicates a contribution of L-tryptophan, which is typically fluorescent at a wavelength longer than that of L-tyrosine. The spectral fitting analysis validated the participation of the two amino acids in the overall emission of the 220 nm band, and also allowed us to estimate their relative contribution. L-tryptophan was indicated to provide a contribution higher than that of L-tyrosine in both biological samples, while the L-tyrosine signal is more than doubled in mosquito eggs compared to the cat black hairs. Two additional spectral components were required to achieve the goodness-of-fit of the procedure. The band peaking at about 310–320 nm is undefined since at the moment, there are no reference data available, while the peaking at about 410–420 nm can be related in general to a proteinaceous material [72].

Interestingly, the UV spectral component relatable to L-tyrosine and L-tryptophan is detected in parallel with the deep red emission still ascribable to these amino acids. For L-tyrosine in particular, this double signature, detected along with the spectra relatable to melanin components, is promising to estimate the degree of melanization process. In this regard, it is worth recalling the reports on the deep red emission detection in various melanin-containing samples of biomedical interest, in particular from skin and hairs, normal nevi and melanomas at different degrees of malignancy [42,43,45–47]. These findings could be accounted for by the influence of the environment and of the excitation technique, namely single-, two- or multiphoton excitation, on the fluorescence of melanin, although the presence of amino acids reflecting the tissue engagement in melanization cannot be disregarded.

In insects, melanin has been suggested to affect desiccation resistance through multiple mechanisms, depending on its covalent or noncovalent interaction with proteins, chitin and other biomolecules [73], or more generally on its hydrophobicity, which hampers the water flux through the cuticle [74]. In the eggs, melanin can decrease their porosity. This effect is particularly relevant for *Ae. albopictus*, whose eggs become dark black in a short time after oviposition and display high desiccation resistance, with a key impact on the geographical dispersal of this global mosquito invader [59]. Consistently, some of the commonly indicated yellow proteins have been demonstrated to participate in the enzyme conversion of dopachrome along the tyrosine-dependent pathway of tanning, favoring the rapid pigmentation and sclerotization of the chorion of *Ae. albopictus* eggs [70,75]. Implications on mosquito reproduction deriving from melanization effects on the cuticle mechanical stiffness may also regard some structures of adult insects. Indeed, the reduced melanization of the male-specific sex combs on the legs of *Drosophila melanogaster* has been shown to affect mating through changes in their structure, resulting in a diminished ability to grasp females before copulation [76].

More generally, it is also worth recalling that components of the melanin biosynthetic pathway have been previously shown to contribute to the wing-scale color of butterflies [51, 77], and that melanization is known to play key roles in the structure and color of the cuticle [78–80].

5. Conclusions

Besides confirming the typical fluorescence spectral properties of L-tyrosine and melanin, variable excitation and emission spectral profiles were collected during the melanization process, providing a basis for the subsequent analysis of samples from black cat hairs and mosquito eggs. The use of these biological samples validated the ability of the mild extraction procedure applied to produce aqueous extracts suitable for the fluorescence analysis. The fluorescence spectra, in turn, were, in general, consistent with those observed from the pure compounds and during the melanization process. The lower signals from the black cat hairs, in turn, could be ascribed to the tight binding of melanin to the proteinaceous components of the hair's mature melanosomes, hindering its extraction compared with mosquito eggs. Improvements in the fluorescence characterization of the melanization process in mosquitoes are expected to contribute to better clarify the coloring of biometabolic pathways, along with the roles of melanin and of its precursors relevant to eggs and cuticle functionality, as well as potential novel functions of fluorescence in mediating intra- and intersexual interactions. This additional optically based information in mosquito species may stimulate the development of novel approaches for tracing and controlling mosquitoes, as well as inspiring novel investigations on the evolutionary meaning of the coloration and fluorescence of eggs and adults.

Supplementary Materials: The following supporting information can be downloaded at: https://www.mdpi.com/article/10.3390/biology12030433/s1, Supplementary text: Supplementary details on the spectral-fitting-analysis procedure; Figure S1: Absorption spectra of L-tyrosine at different times of aging, melanin and L-tryptophan; Figure S2: Fitting of L-tyrosine (A) and L-tryptophan (B); Figure S3: Fitting analysis of emission spectra recorded from black cat hairs (A) and from mosquito eggs (B).

Author Contributions: Conceptualization, methodology, validation, formal analysis, A.C.C.; Investigation, A.C.C. and F.S.; data curation, writing—original draft preparation, visualization, A.C.C.; writing—review and editing, A.C.C. and F.S. All authors have read and agreed to the published version of the manuscript.

Funding: This research received no external funding.

Institutional Review Board Statement: Not applicable.

Informed Consent Statement: Not applicable.

Data Availability Statement: The data presented in this study are available in the article and in the Supplementary Materials file. The raw spectral data are available upon request.

Acknowledgments: The authors acknowledge Giordano Liberi (IGM-CNR, Pavia, Italy) for providing L-tryptophan, as well as Arianna Puggioli and Romeo Bellini (Centro Agricoltura Ambiente "G. Nicoli", Crevalcore, Italy) for providing *Aedes albopictus* eggs.

Conflicts of Interest: The authors declare no conflict of interest.

References

1. Kim, Y.J.; Uyama, H. Tyrosinase inhibitors from natural and synthetic sources: Structure, inhibition mechanism and perspective for the future. *Cell. Mol. Life Sci.* **2005**, *62*, 1707–1723. [CrossRef] [PubMed]
2. Liu, Y.; Hong, L.; Wakamatsu, K.; Ito, S.; Adhyaru, B.; Cheng, C.-Y.; Bowers, C.R.; Simon, J.D. Comparison of structural and chemical properties of black and red human hair melanosomes. *Photochem. Photobiol.* **2005**, *81*, 135–144. [CrossRef]
3. Khayrova, A.; Lopatin, S.; Varlamov, V. Obtaining chitin, chitosan and their melanin complexes from insects. *Int. J. Biol. Macromol.* **2021**, *167*, 1319–1328. [CrossRef]
4. Derby, C.D. Cephalopod ink: Production, chemistry, functions and applications. *Mar. Drugs* **2014**, *12*, 2700. [CrossRef] [PubMed]
5. Palumbo, A. Melanogenesis in the ink gland of *Sepia officinalis*. *Pigment Cell Res.* **2003**, *16*, 517–522. [CrossRef]
6. Slominski, A.; Zmijewski, M.A.; Pawelek, J. L-tyrosine and L-dihydroxyphenylalanine as hormone-like regulators of melanocyte functions. *Pigment Cell Melanoma Res.* **2012**, *25*, 14–27. [CrossRef]
7. Solano, F. Photoprotection and skin pigmentation: Melanin-related molecules and some other new agents obtained from natural sources. *Molecules* **2020**, *25*, 1537. [CrossRef]
8. Hill, H.Z.; Li, W.; Xin, P.; Mitchell, D.L. Melanin: A two edged sword? *Pigment Cell Res.* **1997**, *10*, 158–161. [CrossRef] [PubMed]
9. Nasti, T.H.; Timares, L. MC1R, eumelanin and pheomelanin: Their role in determining the susceptibility to skin cancer. *Photochem. Photobiol.* **2015**, *91*, 188–200. [CrossRef]
10. Felix, C.C.; Hyde, J.S.; Sarna, T.; Sealy, R.C. Interactions of melanin with metal ions. Electron spin resonance evidence for chelate complexes of metal ions with free radicals. *J. Am. Chem. Soc.* **1978**, *100*, 3922–3926. [CrossRef]
11. Zucca, F.A.; Giaveri, G.; Gallorini, M.; Albertini, A.; Toscani, M.; Pezzoli, G.; Lucius, R.; Wilms, H.; Sulzer, D.; Ito, S.; et al. The neuromelanin of human *substantia nigra*: Physiological and pathogenic aspects. *Pigment Cell Res.* **2004**, *17*, 610–617. [CrossRef] [PubMed]
12. Mackintosh, J.A. The antimicrobial properties of melanocytes, melanosomes and melanin and the evolution of black skin. *J. Theor. Biol.* **2001**, *211*, 101–113. [CrossRef]
13. Nappi, A.J.; Christensen, B.M. Melanogenesis and associated cytotoxic reactions: Applications to insect innate immunity. *Insect Biochem. Mol. Biol.* **2005**, *35*, 443–459. [CrossRef]
14. Burkhart, C.G.; Burkhart, C.N. The mole theory: Primary function of melanocytes and melanin may be antimicrobial defense and immunomodulation (not solar protection). *Int. J. Dermatol.* **2005**, *44*, 340–342. [CrossRef]

15. Whitten, M.M.A.; Coates, C.J. Re-evaluation of insect melanogenesis research: Views from the dark side. *Pigment Cell Melanoma Res.* **2017**, *30*, 386–401. [CrossRef]
16. Sugumaran, M.; Barek, H. Critical analysis of the melanogenic pathway in insects and higher animals. *Int. J. Mol. Sci.* **2016**, *17*, 1753. [CrossRef]
17. Sarna, T.; Pilas, B.; Land, E.J.; Truscott, T.G. Interaction of radicals from water radiolysis with melanin. *Biochim. Biophys. Acta* **1986**, *883*, 162–167. [CrossRef]
18. Slominski, R.M.; Sarna, T.; Płonka, P.M.; Raman, C.; Brożyna, A.A.; Slominski, A.T. Melanoma, melanin, and melanogenesis: The yin and yang relationship. *Front. Oncol.* **2022**, *12*, 842496. [CrossRef] [PubMed]
19. Ito, S. A chemist's view of melanogenesis. *Pigment Cell Res.* **2003**, *16*, 230–236. [CrossRef] [PubMed]
20. Bertazzo, A.; Costa, C.V.L.; Allegri, G.; Favretto, D.; Traldi, P. Application of matrix-assisted laser desorption/ionization mass spectrometry to the detection of melanins formed from Dopa and dopamine. *J. Mass Spectrom.* **1999**, *34*, 922–929. [CrossRef]
21. Ito, S.; Wakamatsu, K.; Ozeki, H. Chemical analysis of melanins and its application to the study of the regulation of melanogenesis. *Pigment Cell Res.* **2000**, *13* (Suppl. 8), 103–109. [CrossRef]
22. Ito, S.; Wakamatsu, K. Chemical degradation of melanins: Application to identification of dopamine-melanin. *Pigment Cell Res.* **1998**, *11*, 120–126. [CrossRef] [PubMed]
23. Lerche, C.M.; Olsen, P.; Nissen, C.V.; Philipsen, P.A.; Wulf, H.C. A novel LC-MS/MS method to quantify eumelanin and pheomelanin and their relation to UVR sensitivity—A study on human skin biopsies. *Pigment Cell Melanoma Res.* **2019**, *32*, 809–816. [CrossRef] [PubMed]
24. Liu, Y.; Simon, J.D. Isolation and biophysical studies of natural eumelanins: Applications of imaging technologies and ultrafast spectroscopy. *Pigment Cell Res.* **2003**, *16*, 606–618. [CrossRef]
25. Mostert, A.B. Melanin, the what, the why and the how: An introductory review for materials scientists interested in flexible and versatile polymers. *Polymers* **2021**, *13*, 1670. [CrossRef]
26. Pralea, I.E.; Moldovan, R.C.; Petrache, A.M.; Ilieș, M.; Hegheș, S.C.; Ielciu, I.; Nicoară, R.; Moldovan, M.; Ene, M.; Radu, M.; et al. From extraction to advanced analytical methods: The challenges of melanin analysis. *Int. J. Mol. Sci.* **2019**, *20*, 3943. [CrossRef] [PubMed]
27. Gallas, J.M.; Eisner, M. Fluorescence of melanin-dependence upon excitation wavelength and concentration. *Photochem. Photobiol.* **1987**, *45*, 595–600. [CrossRef]
28. Meredith, P.; Riesz, J. Radiative relaxation quantum yields for synthetic eumelanin. *Photochem. Photobiol.* **2004**, *79*, 211–216. [CrossRef]
29. Nighswander-Rempel, S.P.; Riesz, J.; Gilmore, J.; Bothma, J.P.; Meredith, P. Quantitative fluorescence excitation spectra of synthetic eumelanin. *J. Phys. Chem. B* **2005**, *109*, 20629–20635. [CrossRef]
30. Teale, F.W.; Weber, G. Ultraviolet fluorescence of the aromatic amino acids. *Biochem. J.* **1957**, *65*, 476–482. [CrossRef]
31. Teale, F.W. The ultraviolet fluorescence of proteins in neutral solution. *Biochem. J.* **1960**, *76*, 381–388. [CrossRef] [PubMed]
32. Ghisaidoobe, A.B.T.; Chung, S.J. Intrinsic tryptophan fluorescence in the detection and analysis of proteins: A focus on Förster resonance energy transfer techniques. *Int. J. Mol. Sci.* **2014**, *15*, 22518–22538. [CrossRef]
33. Macías, P.; Pinto, M.C.; Gutiérrez-Mérino, C. Long-wavelength fluorescence of tyrosine and tryptophan solutions. *Biochem. Int.* **1987**, *15*, 961–969. [PubMed]
34. Stark, K.B.; Gallas, J.M.; Zajac, G.W.; Eisner, M.; Golab, J.T. Spectroscopic study and simulation from recent structural models for eumelanin: I. Monomer, dimers. *J. Phys. Chem. B* **2003**, *107*, 3061–3067. [CrossRef]
35. Stark, K.B.; Gallas, J.M.; Zajac, G.W.; Eisner, M.; Golab, J.T. Spectroscopic study and simulation from recent structural models for eumelanin: II. Oligomers. *J. Phys. Chem. B* **2003**, *107*, 11558–11562. [CrossRef]
36. Rzepecki, L.M.; Waite, J.H. Aβ-dehydro-3,4-dihydroxyphenylalanine derivatives: Rate and mechanism of formation. *Arch. Biochem. Biophys.* **1991**, *285*, 27–36. [CrossRef]
37. Lee, B.P.; Dalsin, J.L.; Messersmith, P.B. Synthesis and gelation of DOPA-modified poly(ethylene glycol) hydrogels. *Biomacromolecules* **2002**, *3*, 1038–1047. [CrossRef]
38. Li, Y.; Liu, J.; Wang, Y.; Chan, H.W.; Wang, L.; Chan, W. Mass spectrometric and spectrophotometric analyses reveal an alternative structure and a new formation mechanism for melanin. *Anal. Chem.* **2015**, *87*, 7958–7963. [CrossRef]
39. Gauden, M.; Pezzella, A.; Panzella, L.; Napolitano, A.; D'Ischia, M.; Sundström, V. Ultrafast excited state dynamics of 5,6-dihydroxyindole, a key eumelanin building block: Nonradiative decay mechanism. *J. Phys. Chem. B* **2009**, *113*, 12575–12580. [CrossRef]
40. Nighswander-Rempel, S.P.; Mahadevan, I.B.; Rubinsztein-Dunlop, H.; Meredith, P. Time-resolved and steady-state fluorescence spectroscopy of eumelanin and indolic Polymers. *Photochem. Photobiol.* **2007**, *83*, 1449–1454. [CrossRef]
41. Pigoli, C.; Gibelli, L.R.; Caniatti, M.; Moretti, L.; Sironi, G.; Giudice, C. Bleaching melanin in formalin-fixed and paraffin-embedded melanoma specimens using visible light: A pilot study. *Eur. J. Histochem.* **2019**, *63*, 3071. [CrossRef]
42. Leupold, D.; Szyc, L.; Stankovic, G.; Hofmann, M.; Scholz, M.; Forschner, A. Dermatofluoroscopy is also for redheads a sensitive method of early melanoma detection. *Dermatology* **2020**, *236*, 508–516. [CrossRef]
43. Dimitrow, E.; Riemann, I.; Ehlers, A.; Koehler, M.J.; Norgauer, J.; Elsner, P.; König, K.; Kaatz, M. Spectral fluorescence lifetime detection and selective melanin imaging by multiphoton laser tomography for melanoma diagnosis. *Exp. Dermatol.* **2009**, *18*, 509–515. [CrossRef] [PubMed]

44. König, K.; Ehlers, A.; Stracke, F.; Riemann, I. *In vivo* drug screening in human skin using femtosecond laser multiphoton tomography. *Skin Pharmacol. Physiol.* **2006**, *19*, 78–88. [CrossRef]
45. Teuchner, K.; Ehlert, J.; Freyer, W.; Leupold, D.; Altmeyer, P.; Stücker, M.; Hoffmann, K. Fluorescence studies of melanin by stepwise two-photon femtosecond laser excitation. *J. Fluoresc.* **2000**, *10*, 275–281. [CrossRef]
46. Krasieva, T.B.; Stringari, C.; Liu, F.; Sun, C.-H.; Kong, Y.; Balu, M.; Meyskens, F.L.; Gratton, E.; Tromberg, B.J. Two-photon excited fluorescence lifetime imaging and spectroscopy of melanins In Vitro and In Vivo. *J. Biomed. Opt.* **2013**, *18*, 031107. [CrossRef] [PubMed]
47. Eichhorn, R.; Wessler, G.; Scholz, M.; Leupold, D.; Stankovic, G.; Buder, S.; Stücker, M.; Hoffmann, K. Early diagnosis of melanotic melanoma based on laser-induced melanin fluorescence. *J. Biomed. Opt.* **2009**, *14*, 034033. [CrossRef]
48. Lapierre-Landry, M.; Carroll, J.; Skala, M.C. Imaging retinal melanin: A review of current technologies. *J. Biol. Eng.* **2018**, *12*, 29. [CrossRef]
49. Schmitz-Valckenberg, S.; Pfau, M.; Fleckenstein, M.; Staurenghi, G.; Sparrow, J.R.; Bindewald-Wittich, A.; Spaide, R.F.; Wolf, S.; Sadda, S.R.; Holz, F.G. Fundus autofluorescence imaging. *Prog. Retin Eye Res.* **2021**, *81*, 100893. [CrossRef]
50. Keilhauer, C.N.; Delori, F.C. Near-infrared autofluorescence imaging of the fundus: Visualization of ocular melanin. *Investig. Ophthalmol. Vis. Sci.* **2006**, *47*, 3556–3564. [CrossRef]
51. Matsuoka, Y.; Monteiro, A. Melanin pathway genes regulate color and morphology of butterfly wing scales. *Cell Rep.* **2018**, *24*, 56–65. [CrossRef] [PubMed]
52. Stavenga, D.G.; Leertouwer, H.L.; Hariyama, T.; de Raedt, H.A.; Wilts, B.D. Sexual dichromatism of the damselfly *Calopteryx japonica* caused by a melanin-chitin multilayer in the male wing veins. *PLoS ONE* **2012**, *7*, e49743. [CrossRef] [PubMed]
53. Li, J.; Zhao, X.; Christensen, B.M. Dopachrome conversion activity in *Aedes aegypti*: Significance during melanotic encapsulation of parasites and cuticular tanning. *Insect Biochem. Mol. Biol.* **1994**, *24*, 1043–1049. [CrossRef] [PubMed]
54. Johnson, J.K.; Li, J.; Christensen, B.M. Cloning and characterization of a dopachrome conversion enzyme from the yellow fever mosquito, *Aedes aegypti*. *Insect Biochem. Mol. Biol.* **2001**, *31*, 1125–1135. [CrossRef] [PubMed]
55. Croce, A.C.; Scolari, F. Autofluorescent biomolecules in Diptera: From structure to metabolism and behavior. *Molecules* **2022**, *27*, 4458. [CrossRef] [PubMed]
56. Gebru, A.; Jansson, S.; Ignell, R.; Kirkeby, C.; Prangsma, J.C.; Brydegaard, M. Multiband modulation spectroscopy for the determination of sex and species of mosquitoes in flight. *J. Biophotonics* **2018**, *11*, e201800014. [CrossRef]
57. Nansen, C. Penetration and scattering—Two optical phenomena to consider when applying proximal remote sensing technologies to object classifications. *PLoS ONE* **2019**, *13*, e0204579. [CrossRef] [PubMed]
58. Scolari, F.; Girella, A.; Croce, A.C. Imaging and spectral analysis of autofluorescence patterns in larval head structures of mosquito vectors. *Eur. J. Histochem.* **2022**, *66*, 3462. [CrossRef]
59. Farnesi, L.C.; Vargas, H.C.M.; Valle, D.; Rezende, G.L. Darker eggs of mosquitoes resist more to dry conditions: Melanin enhances serosal cuticle contribution in egg resistance to desiccation in *Aedes*, *Anopheles* and *Culex* vectors. *PLoS Negl. Trop. Dis.* **2017**, *11*, e0006063. [CrossRef]
60. Anderson, P.J.B.; Rogers, Q.R.; Morris, J.G. Cats require more dietary phenylalanine or tyrosine for melanin deposition in hair than for maximal growth. *J. Nutr.* **2002**, *132*, 2037–2042. [CrossRef]
61. Prota, G. *Melanins and Melanogenesis*; Academic Press: San Diego, CA, USA, 1992; p. 290.
62. Pudar, D.; Puggioli, A.; Balestrino, F.; Sy, V.; Carrieri, M.; Bellini, R.; Petrić, D. Effect of cage size on *Aedes albopictus* wing length, survival and egg production. *Heliyon* **2021**, *7*, e07381. [CrossRef]
63. Marquardt, D.W. An Algorithm for least-squares estimation of nonlinear parameters. *SIAM J. Appl. Math.* **1963**, *11*, 431–441. [CrossRef]
64. Hutnik, C.M.L.; Szabo, A.G. Long-wavelength fluorescence of tyrosine and tryptophan: A classic example of second order diffraction. *Biochem. Int.* **1988**, *16*, 587–591. [PubMed]
65. Meredith, P.; Powell, B.J.; Riesz, J.; Nighswander-Rempel, S.P.; Pederson, M.R.; Moore, E.G. Towards structure–property–function relationships for eumelanin. *Soft Matter.* **2006**, *2*, 37–44. [CrossRef]
66. Slawinski, J.; Puzyna, W.; Slawinska, D. Chemiluminescence during photooxidation of melanins and soil humic acids arising from a singlet oxygen mechanism. *Photochem. Photobiol.* **1978**, *28*, 459–463. [CrossRef]
67. Kohl, F.R.; Grieco, C.; Kohler, B. Ultrafast spectral hole burning reveals the distinct chromophores in eumelanin and their common photoresponse. *Chem. Sci.* **2020**, *11*, 1248–1259. [CrossRef] [PubMed]
68. Chen, Y.C.; Kendall, T.; Yip, P.; Davy, A.; Sefcik, J.; Sutter, J.U. Influence of ions and pH on the formation of solid- and liquid-like melanin. *ACS Omega* **2020**, *5*, 25059–25068. [CrossRef]
69. Meng, S.; Kaxiras, E. Theoretical models of eumelanin protomolecules and their optical properties. *Biophys. J.* **2008**, *94*, 2095–2105. [CrossRef]
70. Noh, M.Y.; Mun, S.; Kramer, K.J.; Muthukrishnan, S.; Arakane, Y. Yellow-y Functions in egg melanization and chorion morphology of the Asian tiger mosquito, *Aedes albopictus*. *Front. Cell Dev. Biol.* **2021**, *9*, 769788. [CrossRef]
71. Novellino, L.; Napolitano, A.; Prota, G. Isolation and characterization of mammalian eumelanins from hair and irides. *Biochim. Biophys. Acta Gen. Subj.* **2000**, *1475*, 295–306. [CrossRef]

72. Palladini, G.; di Pasqua, L.G.; Cagna, M.; Croce, A.C.; Perlini, S.; Mannucci, B.; Profumo, A.; Ferrigno, A.; Vairetti, M. MCD diet rat model induces alterations in zinc and iron during NAFLD progression from steatosis to steatohepatitis. *Int. J. Mol. Sci.* **2022**, *23*, 6817. [CrossRef] [PubMed]
73. Arakane, Y.; Noh, M.Y.; Asano, T.; Kramer, K.J. Tyrosine metabolism for insect cuticle pigmentation and sclerotization. In *Extracellular Composite Matrices in Arthropods*; Springer International Publishing: Cham, Switzerland, 2016; pp. 165–220.
74. Rajpurohit, S.; Peterson, L.M.; Orr, A.J.; Marlon, A.J.; Gibbs, A.G. An experimental evolution test of the relationship between melanism and desiccation survival in insects. *PLoS ONE* **2016**, *11*, e0163414. [CrossRef] [PubMed]
75. Noh, M.Y.; Kim, S.H.; Gorman, M.J.; Kramer, K.J.; Muthukrishnan, S.; Arakane, Y. Yellow-g and yellow-G2 proteins are required for egg desiccation resistance and temporal pigmentation in the Asian tiger mosquito, *Aedes albopictus*. *Insect Biochem. Mol. Biol.* **2020**, *122*, 103386. [CrossRef] [PubMed]
76. Massey, J.H.; Chung, D.; Siwanowicz, I.; Stern, D.L.; Wittkopp, P.J. The *yellow* gene influences *Drosophila* male mating success through sex comb melanization. *Elife* **2019**, *8*, e49388. [CrossRef]
77. Ito, K.; Yoshikawa, M.; Fujii, T.; Tabunoki, H.; Yokoyama, T. Melanin pigmentation gives rise to black spots on the wings of the silkworm *Bombyx mori*. *J. Insect Physiol.* **2016**, *91–92*, 100–106. [CrossRef]
78. Vincent, J.F.V.; Wegst, U.G.K. Design and mechanical properties of insect cuticle. *Arthropod. Struct. Dev.* **2004**, *33*, 187–199. [CrossRef]
79. Andersen, S.O. 6—Cuticular sclerotization and tanning. In *Insect Molecular Biology and Biochemistry*; Gilbert, L.I., Ed.; Elsevier: Amsterdam, The Netherlands; Academic Press: Cambridge, MA, USA, 2012; pp. 167–192. [CrossRef]
80. Reinhardt, K.; Breunig, H.G.; König, K. Autofluorescence lifetime variation in the cuticle of the bedbug *Cimex lectularius*. *Arthropod. Struct. Dev.* **2017**, *46*, 56–62. [CrossRef]

biology

MDPI

Article

Evaluation of Thermal Changes of the Sole Surface in Horses with Palmar Foot Pain: A Pilot Study

Cristian Zaha [1,*], Larisa Schuszler [1], Roxana Dascalu [1], Paula Nistor [1], Tiana Florea [2], Kálmán Imre [3], Ciprian Rujescu [4], Bogdan Sicoe [5] and Cornel Igna [1,*]

1 Surgery Clinic, Faculty of Veterinary Medicine, University of Life Sciences "King Michael I", 300645 Timisoara, Romania
2 Dermatology Department, Faculty of Veterinary Medicine, University of Life Sciences "King Mihai I", 300645 Timisoara, Romania
3 Department of Animal Production and Veterinary Public Health, Faculty of Veterinary Medicine, University of Life Sciences "King Michael I", 300645 Timisoara, Romania
4 Management and Rural Development Department, Faculty of Management and Rural Tourism, University of Life Sciences "King Michael I", 300645 Timisoara, Romania
5 Diagnostic Imaging, Faculty of Veterinary Medicine, University of Life Sciences "King Michael I", 300645 Timisoara, Romania
* Correspondence: cristian.zaha@usab-tm.ro (C.Z.); ignacornel@gmail.com (C.I.)

Simple Summary: A pilot study was conducted to investigate the accuracy of thermography used to detect changes in the local temperature of horse feet and to compare the thermal patterns observed on the sole surface after training. The study group included eight horses with unilateral forelimb palmar foot pain (n = 8) and healthy contralateral limbs (n = 8). Four additional non-lame horses served as controls. We compared the local temperature of the frog and toe areas of the limbs affected with palmar foot pain in the study group with the other non-lame limbs in the study group and control group. The temperatures of the two selected areas were considered when determining differences between the three groups of horse limbs. The temperature of the frog area did not increase, and its area did not expand after training in the limbs affected with palmar foot pain. The temperature of the sole and its area both increased after training among the affected limbs of horses with palmar foot pain compared to the non-lame limbs. Based on the results obtained in this pilot study, we suggest a thermographic scanning of the toe and frog area to detect changes in the local temperature of the sole surface after training to discriminate the horses with palmar foot pain from non-lame ones.

Abstract: Background: Horses with palmar foot pain do not show a typical increase in temperature in the palmar aspect of the hoof and heel due to low blood flow. The objectives of the current study were to determine the changes and differences in the thermographic pattern of the sole surface in horses with unilateral palmar foot pain and non-lame horses before and after training. We hypothesized that the thermal pattern and the local temperature of the frog area and toe area would be similar, with an increased local temperature after training in both lame and non-lame horses. A pilot study was conducted to investigate the accuracy of thermography used to detect changes in local temperature and to compare the thermal patterns observed on the sole surface after training. Methods: The study group included eight horses with unilateral forelimb palmar foot pain (n = 8) and healthy contralateral limbs (n = 8). Four additional non-lame horses served as controls. The horses were selected for the study based on the following criteria: forelimb with unilateral chronic progressive lameness and positive reactions when performing the hoof test and frog wedge test, degenerative findings of the navicular bone, and improvement in lameness after perineural analgesia of the medial and lateral palmar digital nerve. The local temperatures of the frog and toe areas were measured before and after training in the affected and contralateral limbs in the study group and both forelimbs in the control group using an FLIR E50 thermal camera. Receiver operating characteristic (ROC) curve analysis was applied to check the sensitivity and specificity of the results for the studied groups. Results: The thermal patterns of the hoof presented differences between the three groups of horse limbs. After training, the temperature of the sole surface increased, and its area increased in the limbs affected with palmar foot pain compared with the non-lame limbs in the study group and the limbs in the control group. The temperature of the frog area did not increase after training in the limbs affected with palmar foot pain compared with the same area in the other groups. The ROC curve analysis demonstrated the medical applicability of this tool and showed that thermography is a reliable diagnostic test to correctly discriminate between non-lame horses and those with palmar foot pain. Conclusions: We recommend thermographic scanning of the toe and frog area to detect changes in the local temperature of the sole surface after training to discriminate the horses with palmar foot pain from non-lame ones. Further investigation is required to clarify whether the observed thermographic imaging features of the sole surface are characteristic for horses with palmar foot pain.

Keywords: lameness; palmar foot pain; thermography

1. Introduction

Palmar foot pain represents a common cause of lameness in horses and can be the result of navicular disease, navicular bursitis, desmitis to the distal sesamoidean impar ligament (DSIL), desmitis of the collateral ligaments of the distal interphalangeal joint (DIP joint), tendinitis of the deep digital flexor tendon, laminar tearing or bruising on the palmar region, distal phalanx damage, subsolar seromas, and abscesses [1–3]. Horses aged between 6 and 10 years are most commonly affected [4]. The etiology of palmar foot pain remains unknown, but two main theories, abnormal biomechanical stresses, and vascular compromise, have been suggested [5,6].

Diagnoses of palmar foot pain are based on history, physical examination, lameness examination, hoof testing, and perineural and/or intra-articular diagnostic anesthesia. Several imaging techniques (e.g., radiography (Rx), ultrasonography (US), nuclear scintigraphy (SCI), thermography (IRT), computed tomography (CT), magnetic resonance imaging (MRI)) are subsequently performed to ensure accurate diagnosis and proper treatment [7–9]. Turner [10] stated that in order to exacerbate the pressure on the navicular apparatus, a wooden pad should be placed under the frog area and the animal forced to land on it; this test is known as the frog wedge test.

Thermography is a non-invasive diagnostic method that involves scanning the surface temperature of an animal to find areas presenting potential vascular changes [11,12]. This ability to non-invasively assess vascular changes makes thermography an ideal imaging tool to aid in the diagnosis of certain lameness conditions in horses [13]. Thermography can support additional diagnostic information, together with Rx, US, CT, to detect vascular changes associated with hoof pain, such as laminitis, sole abscesses, navicular disease, and stress fracture, and, for the thoracolumbar region, it is used to scan the spinous processes of the thoracic vertebrae for diagnosis of inflammation and myositis of the region [14,15]. In thermographic imaging, the variation in the color pattern reflects thermal gradients; thus, the warmer areas with increased blood circulation are depicted as being white or red, while the cooler regions with insufficient blood supply appear blue or black [13]. Many studies emphasize the importance of thermography as a complement to ultrasonographic and radiographic examinations in the evaluation of soft tissue injuries and superficial bone lesions [12,14,15]. The highest body-surface temperatures (27–32 °C) are recorded in areas such as the flank, shoulders, head, and neck, while the lowest temperatures (24–26 °C) are recorded in the distal limbs [16]. The coronary band is 1 °C to 2 °C warmer than the remainder of the hoof [13].

Because thermography is non-invasive and inexpensive, it may be used to detect early changes prior to lameness occurring [17].

Turner et al. evaluated the local temperature of the caudal part of the foot to compare differences between non-lame horses and those suffering from navicular syndrome; horses with palmar foot pain did not display a typical increase in temperature in the palmar aspect of the hoof skin due to low blood flow after training [18].

The objectives of the current study were as follows:

1. To determine changes in the thermographic pattern of the sole surface in horses with unilateral palmar foot pain before and after training;
2. To assess the differences in the thermal pattern of the sole in horses with palmar foot pain and non-lame horses before and after training;
3. To verify that thermographic examination can be used to detect changes in the local temperature of the sole before and after training.

We hypothesized that the thermal pattern and the local temperature of the frog and toe area will not undergo temperature changes after training in the examined areas in any of the three groups.

2. Materials and Methods

2.1. Animal Selection

A prospective study was performed at the Surgery Clinic of the Faculty of Veterinary Medicine of Timisoara during the autumn season of 2021. Eight horses with unilateral forelimb lameness and four non-lame horses were selected. The horses' feet were divided into the following three groups: group one (study group) represented by the affected forelimbs (n = 8), group two (study group) represented by the contralateral non-lame forelimbs (n = 8), and group three (control group) represented by the non-lame limbs of the non-lame horses (n = 4).

The inclusion criteria for the study group were as follows: more than three months of progressive unilateral forelimb lameness, minimum lameness grade of 2/10 on straight line, positive reactions to the hoof testers applied on the frog area and frog wedge pressure test, soundness after bilateral digital palmar perineural analgesia, radiographic observations, such as enthesophytes, on the medial and lateral aspects of the navicular bone, increased medullary sclerosis, and enlarged synovial invaginations [19,20]. No anti-inflammatory medication was administrated to the horses in the three weeks before the examination. The exclusion criteria for the study group were tenosynovitis and arthritis/arthrosis of the metacarpo-phalangeal joint.

The horses were clinically examined by two veterinarians with 10 and 14 years of experience, respectively, in musculoskeletal injuries, with a lameness exam performed according to the subjective European scale of lameness (0 is sound; 10 is non-weight-bearing lameness) [21]. The horses started to become lame while trotting on a hard surface, and this was even more obvious in a circle with the affected limb on the inside. The diameter of the circle around which the horses were lunged was 15 m. Lameness was obvious on the straight line in three horses and became worse during the lunge with the affected limb on the interior of the circle.

The hoof tester was applied on the frog area by squeezing the frog with one hand and the hoof wall with the other hand, to exert pressure on the navicular apparatus. The frog wedge test was performed by applying a wooden pad that was 10 cm long and 5 cm wide, with an angle of 15 degrees, under the palmar two thirds of the frog and forcing the horse to stand on this foot for 1 min by lifting the contralateral limb. The horses were walked after the frog test and a positive response was recorded if the lameness worsened.

To localize the lameness, perineural analgesia of the medial and lateral palmar digital nerve was performed in the pastern area at 1 cm proximal to the proximal edge of the collateral cartilage [17,22–24]. Before administration of the anesthetic solution, the hair was clipped, and the skin was scrubbed with povidone iodine. Perineural analgesia of the medial and lateral digital palmar nerve was performed using a 1.5-milliliter solution of 2% mepivacaine (Scandicaine®) for each nerve, and the size of the needle was 25 gauge and 16 mm [25]. Thirty minutes after perineural analgesia, the horses were subjected to a lameness examination [26].

Radiographic examination included the following views: lateromedial (LM), 65° dorso proximal–palmarodistal oblique (DPr-PaDiO), and 45° palmaroproximal–palmarodistal oblique (PaPr-PaDIO) [19,20].

The control group was subjected to the same clinical exam, lameness and pressure tests, and radiographic imaging. For the control group, the lateral and medial digital nerve did not receive perineural analgesia. The exclusion criteria for the control group were no anti-inflammatory drugs and no history of tenosynovitis or hoof abscess in the last month.

2.2. Thermal Imaging and Data Recording

We evaluated the intensity of radiation in the infrared spectrum of the toe and frog area (Figure 1) before and after a 30-min lunging training session in each group. A qualitative assessment for the hoof wall area was performed before and after the training session. All the horses were subjected to the same intensity of exercise: 15-min walk on left and right hand and 15-min trot on left and right hand. The frog area was named El1, located in the middle third of the frog between the para-cuneal grooves at a size of 8 × 4 pixels. The toe area was named El2, located between the apex of the frog and hoof wall at a size of 6 × 4 pixels. The training area comprised a paddock built from quartz sand and the horses were lunged on both the right- and left-hand sides. The paddock surface was dry during the training sessions.

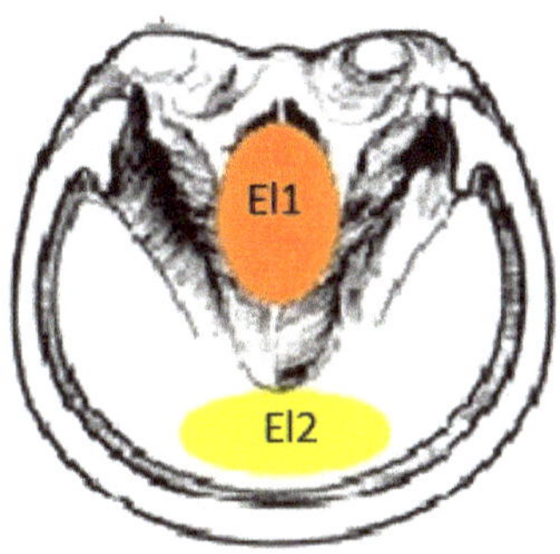

Figure 1. Sole surface: El1, frog area; El2, toe area.

A thermographic exam was performed the day after the clinical exam and perineural analgesia.

All measurements for each horse in the study were conducted during the same season in the same environmental conditions, i.e., dry sand, air temperature of 18–22 °C, humidity between 60% and 70%, without extreme heat or cold weather. The acclimation period for thermal scanning was 30 min before training in a closed space with a temperature of 20–22 °C; after training, we did not include an acclimation period but proceeded directly with scanning [27]. The thermal images and measurements were taken by the same non-blinded operator for non-lame and lame horses at a 50 cm distance and 90° angle to the sole surface. Images were obtained using a FLIR E50 (FLIR Systems Inc., Wilsonville, OR, USA) infrared camera with the following parameters: 0.95 emissivity and 240 × 180 resolution for each image. The temperature range was between −20 and 650 °C, and the sensitivity was ≤0.05 °C. To identify the minimum and mean temperatures of the projection, we used FLIR Tools software for photo interpretation and the procedure was performed by the same operator.

For horses from the study group, the thermographic evaluation was performed between 10 a.m. and 12 p.m. on different days of the autumn season without differences in the outside temperature. For horses from the control group, the thermographic evaluation was performed in the same autumn season and on the same day so that the outside temperature did not influence the obtained results.

2.3. Statistical Analysis

The Kruskal–Wallis statistical test was applied to the temperature, and four procedures were used to evaluate each group. Comparisons were made between the control group, study group with the non-lame limb, and study group with palmar foot pain for the following four indicators (minimum temperature before training; minimum temperature after training; mean temperature before training; mean temperature after training) in both the toe and frog area. The Kruskal–Wallis statistical test was applied using IBM SPSS Statistics software version 23.0 (IBM, Armonk, NY, USA) with the statistical significance set at $p < 0.05$. Pairwise comparisons were also performed between the indicated groups using Dunn's post hoc test with Bonferroni adjustment.

The receiver operating characteristic (ROC) curve is created from the sensitivity and specificity data and allows the creation of cut-off values. The purpose of ROC analysis is to identify the optimal threshold value to differentiate a positive from a negative result by obtaining the value of the area under the curve (AUC). To determine the ROC curve for toe area, the data were processed following attaching for each specimen of the group a score (temperature), respectively, which was a binary value (0—without lameness, 1—with lameness). To determine the ROC curve for frog area, we took into account the fact that in study group with palmar foot pain the temperature after training does not undergo obvious changes, the score attaching for each specimen was calculated as 1/t, where 1/t represent the inversion temperature for minimum and mean temperature before and after training. The PSPP software was used to obtain the results for ROC curve (PSPP – Analyze - ROC Curve). If the obtained AUC is greater than 0.9, then the result is excellent. If the obtained area is between 0.8 and 0.9, the result is very good; if it is between 0.7 and 0.8, the result is good; if it is between 0.6 and 0.7, the result is fair; and if it is less than 0.6, the result is dismissed.

The mean value for the minimum temperature prior to and subsequently after training was calculated separately for each group. The element of comparison of the evolution of the temperature after the training session between groups was the average value obtained from the mean of the minimum values of temperatures with the abbreviation of m.m.v.t.

3. Results

3.1. Animal Information

Out of fifty-eight horses examined during the autumn season of 2021 in the Surgery Clinic of Faculty of Veterinary Medicine from Timisoara, twelve were included in the study, and from these, four were sound horses and eight horses presented with unilateral forelimb lameness. All horses were Warmblood horses; three of the horses were used for cross-country sports and nine were used for leisure. The horses ranged in age from 6 to 14 years (mean 10.25 years and median 10 years). Among the horses from the study group, two presented with slight lameness on walking, and the other six had a history of lameness after exercise. Five horses were shod and three horses were bare-hoofed, and the horseshoes were removed from all horses before examination. After shoe removal, the hooves were trimmed and adjusted by a qualified farrier with more than 15 years of experience in horseshoe and hoof balance for racehorses. A control group composed of four barefoot, non-lame horses was also included in the study ($n = 4$). The horses from the control group were barefoot and a single forelimb was taken into account. These horses had private owners and were brought to the clinic for routine checks; they participated in the study with the consent of the owners. During the clinic examination, none of the horses presented with any musculoskeletal conditions.

3.2. Lameness Exam

Three horses were lame on the left forelimb and five were lame on the right forelimb; these ones represent the study group with palmar foot pain. All horses had a positive response to pain after flexion of the distal forelimb, which produced a transient exacerbation of the lameness. A lameness score of 4 was found in two of the studied horses, a score of 6 was found in three horses, and 7 in three horses, considering the European scale for lameness.

Six of the horses presented a mild reaction and two horses had a moderate reaction of retraction after applying the hoof tester on the frog area of the hoof. After using the wedge frog test, three of the horses presented a mild reaction, three horses a moderate reaction, and two horses an obvious reaction.

After perineural analgesia of the medial and lateral digital nerve, the lameness score was reduced to 0 in four horses (before: scores of 4 and 7), a score of 1 for three horses (before: scores of 6 and 4), and a score of 2 for two horses (before: scores of 6 and 7). After perineural analgesia, the response to the hoof tester on the palmar part of the hoof was negative in six horses, and two horses presented a mild reaction (Figure S1).

3.3. Radiographic Exam

In addition, 65° DPr-PaDiO radiographs of the hooves were taken for each of the horses included in the study group, revealing enthesophytes on the medial and lateral aspects of the navicular bone in three of the horses, along with increased medullary sclerosis of the navicular bone in four horses (Figure 2), while PaPr-PaDIO radiographs revealed sclerosis of the medullar cavity in two horses and increased opacity on the flexor border of the navicular bone for three horses. The LM revealed erosions of the flexor cortex of the navicular bone for two horses. Multiple lesions were observed in the case of six horses, and in only two horses did we observe a singular lesion [20]. The radiographic images were read by a veterinarian radiologist with 7 years of experience in the equine field.

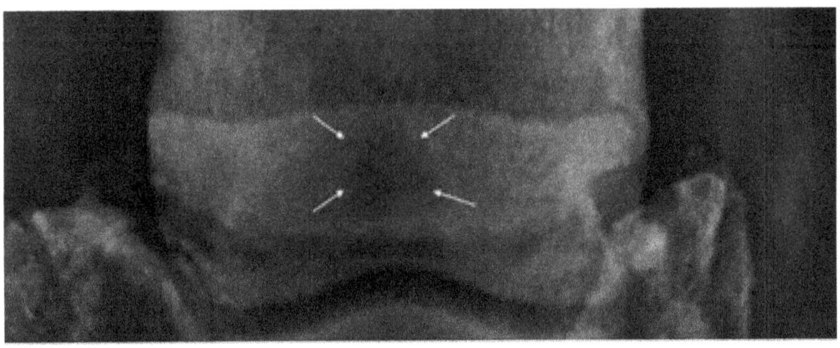

Figure 2. Horses with palmar foot pain: DPr-PaDiO radiolucent area (sclerosis of the medullary cavity) in the navicular bone (white arrows).

The horses from the control group did not present any history of lameness and demonstrated no pain reaction after distal limb flexion. The clinical examination revealed no sign of lameness after walking, trotting, and lunging. The horses did not present any reaction to pain after applying the hoof tester and frog wedge test. The radiographic images did not reveal any changes in the structure of the navicular bone or abnormal opacity in the articular and flexor borders in the LM view. The 65° DPr-PaDiO radiographs of the hooves did not show any enthesophytes or arthrosis of the DIP, and the 45° PaPr-PaDIO radiographs did not detect sclerosis of the medullary cavity of the navicular bone.

3.4. Thermography Scan

3.4.1. Thermography before Training

Thermography of the Toe Area

A small area of increased temperature was noticed between the frog and the hoof wall, which corresponds to the toe area, in the study group with palmar foot pain (Figure 3a and Table 1). The m.m.v.t. of the toe area was 20.6 °C.

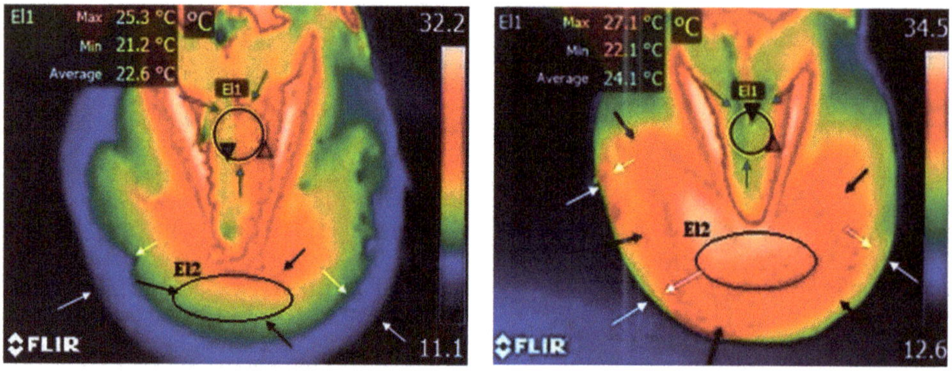

(a) (b)

Figure 3. Study group, limbs with palmar foot pain: (**a**) sole surface before training—small surface of increased temperature in the frog area (blue arrows) and toe area (black arrows); reduced temperature along the hoof wall (white arrows); minimum temperature recorded in the area—blue triangle spot; maximum temperature recorded in the area—red triangle spot; (**b**) sole surface after training—increased temperature and wider surface of increased temperature in the toe area (black arrows); no obvious changes in local temperature in the frog area (blue arrows); increased temperature along the hoof wall (white arrows); minimum temperature recorded in the area—blue triangle spot; maximum temperature recorded in the area—red triangle spot.

Table 1. Descriptive statistics for minimum, maximum, and mean temperature for each group.

				Minimum	Maximum	Median
Control group	frog area	minimum temperature	before	22.3	23.9	22.65
			after	24.1	25.5	24.70
		mean temperature	before	24.7	25.8	25.15
			after	26.1	27.4	26.45
	toe area	minimum temperature	before	22.5	25.1	24.30
			after	25.6	27.2	26.50
		mean temperature	before	25.1	27.4	25.90
			after	27.2	28.0	27.70
Study group, non-lame limb	frog area	minimum temperature	before	25.5	28.5	26.80
			after	27.8	30.8	28.40
		mean temperature	before	27.3	29.9	27.95
			after	28.8	31.9	29.35
	toe area	minimum temperature	before	26.9	28.5	27.55
			after	28.2	29.6	28.90
		mean temperature	before	27.3	29.3	28.35
			after	28.9	30.2	29.55
Study group, palmar foot pain limb	frog area	minimum temperature	before	19.8	23.6	21.20
			after	20.0	24.8	22.85
		mean temperature	before	20.9	24.6	22.95
			after	21.7	26.2	25.55
	toe area	minimum temperature	before	19.5	21.4	20.50
			after	24.2	28.9	27.70
		mean temperature	before	22.8	25.2	23.90
			after	28.7	31.7	29.30

A small area of increased temperature was present near the toe in the study group with the contralateral non-lame limb (Figure 4a and Table 1). The m.m.v.t. value of the toe area was 27.7 °C.

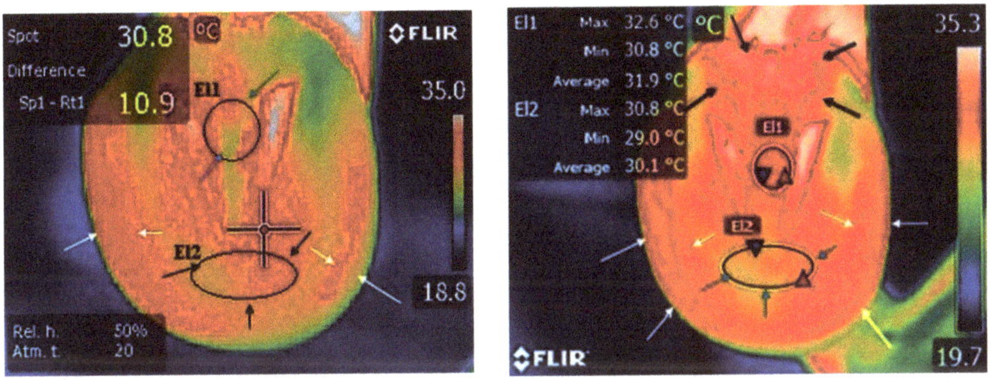

(**a**) (**b**)

Figure 4. Study group, non-lame limbs: (**a**) sole surface before training—reduced temperature and small surface of temperature in the frog area (blue arrows); increased area of temperature on hoof wall surface (white arrows); reduced temperature in the toe area (black arrows); (**b**) sole surface after training—increased temperature and wider surface of increased temperature in the frog area (black arrows); increased surface of temperature along the hoof wall (white arrows); small surface of reduced temperature in toe area (blue arrows); minimum temperature recorded in the area—blue triangle spot; maximum temperature recorded in the area—red triangle spot.

In the case of the control group, the toe area presented limited temperature increases (Figure 5a and Table 1). The m.m.v.t. for toe area was 24.0 °C.

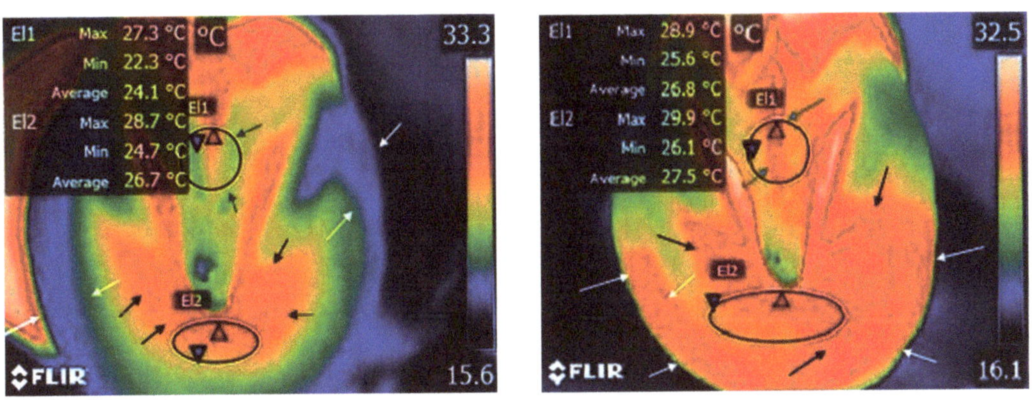

(**a**) (**b**)

Figure 5. Control group: (**a**) small surface of increased temperature in the toe area (black arrows); small surface of increased temperature in the frog area (blue arrows); reduced temperature of the hoof wall (white arrows); minimum temperature recorded in the area—blue triangle spot; maximum temperature recorded in the area—red triangle spot; (**b**) increased local temperature and wider surface of increased temperature in the toe area (black arrows); increased local temperature and wider surface of increased temperature in the frog area (blue arrows); increased temperature along the hoof wall (white arrows); minimum temperature recorded in the area—blue triangle spot; maximum temperature recorded in the area—red triangle spot.

Thermography of the Frog Area

In horses with palmar foot pain, the surrounding temperature was not raised in the frog area (Figure 3a and Table 1), and the m.m.v.t. in the frog area was 21.1 °C.

A small area of increased temperature was present near the frog area in the study group with the contralateral non-lame limb (Figure 4a and Table 1), with the m.m.v.t. in the frog area being 26.7 °C.

In the case of the control group, the frog area was characterized by a decrease in temperature (Figure 5a and Table 1), with the m.m.v.t. of the frog area being 22.8 °C.

3.4.2. Thermography after Training

Thermography of Toe Area

A larger area of increased temperature was found in the toe region in horses with palmar foot pain (Figure 3b, Tables 1 and S1). The m.m.v.t. for the toe area was 27.7 °C.

In the study group with the contralateral non-lame limb, the local temperature of the toe area remained the same (Figure 4a, Tables 1 and S2). The m.m.v.t. for toe area was 28.8 °C.

In the control group the temperature rates in the toe area remained unchanged (Figure 5a, Tables 1 and S3), with m.m.v.t. of 26 °C.

The m.m.v.t. for the toe area reflected an obvious increase in temperature following training sessions, with values of up to 7.1 °C in the case of the palmar foot pain group, respectively, and lower rates for the non-lame limb group, with temperatures increased by only 1.1 °C. Similar to the latter, the temperature increased by only 2 °C in the case of the control group.

After training sessions, the thermal pattern seen in animals from the study group with palmar foot pain was made up of a wider area of increased temperature in the toe region, compared to the same area in the others non-lame groups, with $p = 0.005$.

Thermography of the Frog Area

In the study group with palmar foot pain, a smaller area of lowered temperature was found in the frog area (Figure 3b, Tables 1 and S4), with m.m.v.t of the frog at 22.2 °C.

In the study group with the contralateral non-lame limb, the area of increased temperature widened in the frog area (Figure 4b, Tables 1 and S5), with the m.m.v.t. being 28.7 °C.

In the control group, the frog area presented a wider area of increased temperature rates (Figure 5b, Tables 1 and S6), with the m.m.v.t. of the frog area at 25.2 °C.

The m.m.v.t. was low in the frog area for horses from the study group with palmar foot pain compared to values obtained for the horses from the contralateral non-lame limb and control groups, both before and after training sessions. The m.m.v.t. after training in horses with palmar foot pain increased by 1.1 °C compared to horses from the study group with non-lame limb, where a 2 °C temperature increase was recorded, and to horses from the control group, which showed an increase by 2.4 °C.

The thermal pattern of the frog area is characterized by decreased temperature rates in the case of the study group with the contralateral non-lame limb, compared to the same area in animals from the other non-lame horses, with $p < 0.001$.

Thermography of the Hoof Wall—Qualitative Assessment

In horses with palmar foot pain, before training along the hoof wall, there was a drop in temperature, and after the training, an increased area of high temperature can be seen along the hoof wall (Figure 3).

In the study group with the contralateral non-lame limb, before training, an area of decreased temperature was detected along the hoof wall compared with a widened area of increased temperature after training (Figure 4).

In the control group, an area of decreased temperature was detected along the hoof wall before training compared with a wider area of increased temperature detected after the training (Figure 5).

Before training, based on the thermal pattern, an area of increased temperature was detected along the hoof wall in the study group with the contralateral non-lame limb, compared to the same area for the horses with palmar foot pain and those from the control group.

After training, based on thermal pattern examination, the hoof wall temperature was increased over a wider area in the control group and in the study group with non-lame limbs, as compared to the study group with palmar foot pain.

3.5. Group Comparisons

The result obtained after applying the Kruskal–Wallis test show the $p < 0.001$ after comparison did not differ between the three groups (control group, study group non-lame limb, study group with palmar foot pain) successively following the four indicators for temperature for the toe and frog area. The exceptions for this comparison between the groups were that the minimum temperature for the toe after training had a significance level $p = 0.005$, and the mean temperature of the toe before training were the significance level was $p = 0.01$.

Pairwise comparisons led to significant differences for the frog area after training between the study group limbs with palmar foot pain and the control group ($p < 0.001$) (Figure 6) after Dunn's post hoc test with Bonferroni adjustment. Pairwise comparisons led to a significant difference being found between the study group limbs with palmar foot pain and the study group non-lame limbs regarding the minimum and mean temperatures in the toe area before training ($p < 0.001$). Significant differences were also determined for the control group and study group non-lame limbs regarding the minimum temperature of the toe area after training ($p = 0.002$), with increased temperatures for the study group non-lame limbs. Significant differences were also determined for the mean temperature of the toe area after training between the control group and others study groups ($p = 0.006$) (Figure 7).

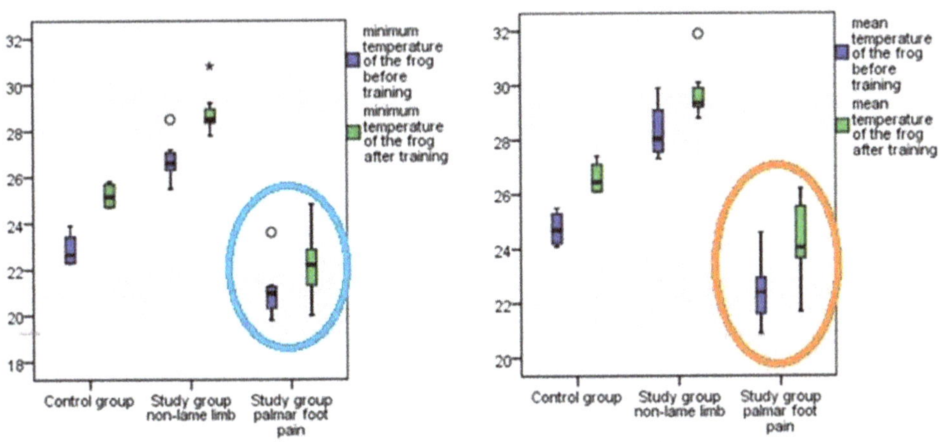

(a) (b)

Figure 6. (**a**) Boxplot diagram showing the minimum temperature values of the frog area before and after training. Low values for minimum temperature before and after training are observed for the study group with palmar foot pain (blue circle). (**b**) Boxplot diagram showing the mean temperature values of the frog area before and after training. Low values for mean temperature before and after training for the study group with palmar foot pain are observed (orange circle); *—extreme outliers temperature, °—mild outliers temperature.

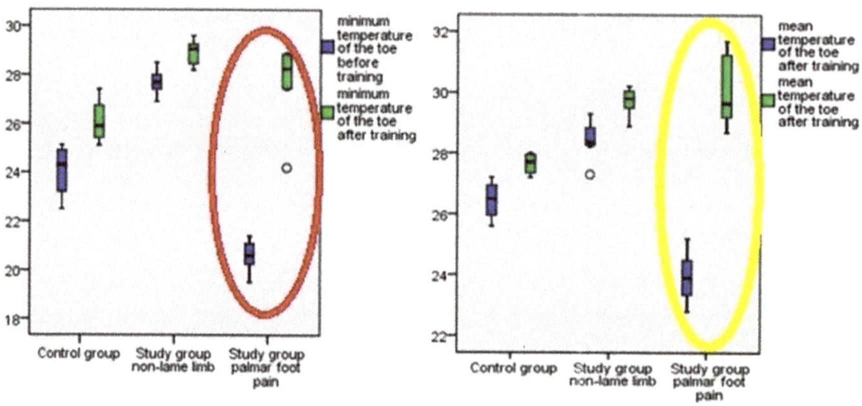

Figure 7. (**a**) Boxplot diagram showing the minimum temperature values of the toe area before and after training. An obvious difference is observed for the minimum temperature of the toe area before (blue mark) and after training (green mark) for the study group with palmar foot pain (red circle). (**b**) Boxplot diagram showing the mean temperature values of the toe area before and after training. Low values for mean temperature of the toe area are observed before (blue mark) and after training (green mark) for the study group with palmar foot pain (yellow circle); °—mild outliers temperature.

3.6. ROC Curve

The ROC curve for the inversion minimum temperature of the frog area before training was 0.97 and $p < 0.05$. The ROC curve for the inversion minimum temperature after training and for the mean temperature before and after training gave a value of the AUC of 0.98 and $p < 0.05$ (Figures 8a and S2).

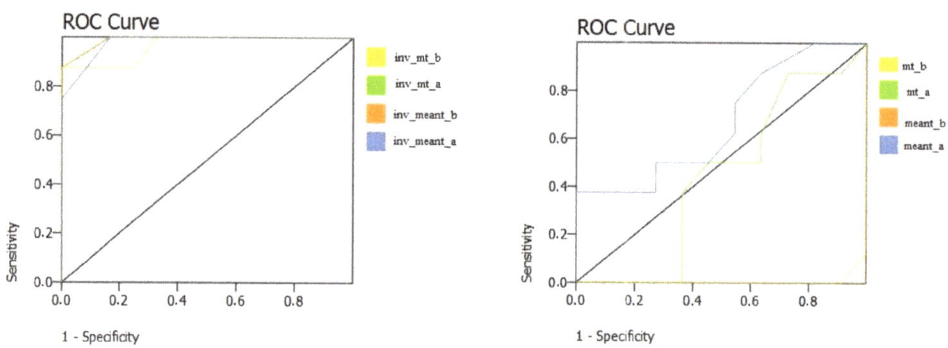

Figure 8. (**a**) Area under the curve (AUC) of the receiver operating characteristic (ROC) curve for frog surface showed the next values for the variants: inv_mt_b—inversion minimum temperature before training—value: 0.97; inv_mt_a—inversion minimum temperature after training—value: 0.98; inv_meant_b—inversion mean temperature before training—value: 0.98; inv_meant_a— inversion mean temperature after training—value: 0.98. (**b**) Area under the curve (AUC) of the receiver operating characteristic (ROC) curve for toe area showed differences between the variants: mt_b—minimum temperature before training—value 0; mt_a—minimum temperature after training value:0.4; meant_b—mean temperature before training—value: 0; meant_a—mean temperature after training value: 0.7.

The ROC curve revealed, for the toe area, two results with an AUC of 0 for the minimum and mean temperature before training, one result with an AUC of 0.7 for the mean temperature of the toe after training, and one result with an AUC of 0.5 and $p > 0.23$ for the minimum temperature of the toe area before training (Figures 8b and S3).

4. Discussion

Thermography of the sole before and after the training session allowed us to differentiate horses with palmar foot pain from non-lame ones; this is highlighted by the rejection of the null hypothesis regarding the minimum and mean temperature before and after training for the frog and toe area. The post-training thermal images from the control group, from the study group with palmar foot pain, and from the study group with non-lame limbs revealed that the local temperature and the thermal pattern of the sole undergo changes after training. The most significant changes in temperature were noticed in the frog and toe areas. The thermal pattern of the toe area after training indicates a significantly larger area of increased temperature in the toe region for the horses with palmar foot pain, as compared to the thermal pattern for the same area in non-lame horses. In addiiton, in horses with palmar foot pain, following thermal scanning after training, a smaller area of lowered temperature in the frog area was observed, as compared to the same area for non-lame horses.

Soroko and Howell report that in horses suffering from disorders affecting the navicular area, the animal is forced to land on the toe instead of the heel, resulting in stride-shortening. The thermographic examination revealed an intensification of blood flow in the sole area, supporting that these animals land improperly on their toes [28]. Our study rendered similar results for horses with palmar foot pain after training, with increased temperatures in the toe area, as to avoid pain in the palmar part of the hoof. The spot showing an increased temperature, identified using thermography, could be differentiated from the increased local temperature linked to the presence of an abscess by associating the pathology with the history of the animal and hoof radiography [29–31].

In our study, the differences in the minimum and mean temperature in the toe area between the control group, non-lame limbs study group, and limbs with palmar foot pain study group appeared because the horses with palmar foot pain were forced to land on the ground on the toe region while horses from the non-lame limbs group were forced to sustain their weight during movements. Thus, in the study group with palmar foot pain, the toe-area showing increased temperature became wider after training, comparable with the same area from the other groups. Buchner et al. indicates a compensation of the reduced loading of the lame limb by the contralateral non-lame limb [32]. This relationship is also visible during lameness, when reaction forces have a tendency to decrease in the lame limb and increase in the contralateral non-lame limb in order to compensate the dysfunction [33].

The pre-training temperature of the frog area in horses with palmar foot pain was decreased compared to the local temperature of the frog area in the other groups of horses. After training, the temperature in the frog area increased by 1.1 °C, on average, in horses with palmar foot pain. As for the other group of horses, the temperature increased by more than 2 °C. Turner [18] reports that in the case of horses with palmar foot pain, around 50% of them do not sustain an increase in temperature in the caudal foot because of low blood flow. This is in contrast with other inflammatory processes of the hoof, such as abscesses, bruises, and fractures, which are characterized by the presence of a "hot spot" at the site of injury, which increases in temperature in the case of exercise [13,18]. A warm spot indicates inflammation or increased circulation, while a cold spot reflects a reduction in blood supply, usually due to swelling, thrombosis, or scar tissue [29,30].

The temperature of the hoof wall increased in value after training for all three groups as a result of increased blood flow in the hooves with consecutive training [34]. Before training, a wider area of temperature was detected for the study group non-lame limbs compared with the same area from other groups as a result of compensatory vertical load bearing, which has the potential to lead to a secondary injury in non-lame horses [35].

All the horses from the studied group with palmar foot pain presented moderate or severe lameness after the clinical examination, a result that other authors have also revealed in their studies of navicular disease and palmar foot pain [6,8,24].

According to the ROC curve, which gave excellent results for three of the eight examined variables with an AUC of 0.98, one with AUC of 0.97, and one with AUC of 0.7, thermography provides excellent clinical utility to correctly identify the horses with palmar foot pain and non-lame ones. For three variables, the value of the AUC was less than 6 (two results AUC of 0, one result AUC of 0.5), so the result was dismissed. As a result, for AUC values, we recommended thermographic scanning of the toe and frog area to claim changes in temperature for horses with palmar foot pain. Compared to other diagnostic imaging procedures (US, Rx, CT, MRI), thermography is inexpensive, requires no harmful radiation, carries no risk of injury to the veterinarian, and does not require the horses to be sedated [11,12,17].

To obtain a successful thermal image the horses must have a dry coat and skin, without sweating, should not have performed physical therapy in the past 24 h, and no anesthetic protocol should be used [15,33,36]. The thermal scanning can be performed within 45 min of perineural anesthesia, without the risk of changes in the local temperature of the limb surface [37]. In our study, the thermography was performed on a dry sole, without artefacts, sweating, the next day after the lameness examination, perineural anesthesia, and without general sedation.

During exercise, working muscles produce substantial quantities of heat, which must be eliminated from the body by sweating, in order to prevent the animal from overheating [18]. In our study, sweating does not appear to influence the thermographic images due to a lack of intense exercises and due to the absence of the eccrine gland on the surface of the frog [38,39].

Our recommendation regarding the evaluation of the sole surface local temperature before and after training is to perform the thermography on a dry area, without artefacts, in an environment without extreme temperature variations and to perform the training session on a dry surface.

The obtained results indicate a disturbance of the local temperature in the toe region in horses with palmar foot pain, and we suggest that this increase in toe temperature may be related to differences in loading characteristics of the foot. Further investigations based on sole surface thermography are required. These should aim to: (1) assess the correlation between biomechanics changes and sole thermography in palmar foot pain horses; (2) compare the thermal pattern areas for non-lame and affected horses before and after training using specific software for area measurement in the SI units (International System of Units); (3) discover whether there is a difference in the thermal pattern in horses with bilateral palmar foot pain and those with unilateral palmar foot pain; and (4) identify the differences between horses with subclinical palmar foot pain and non-lame horses.

The limitations of the study were the small number of studied cases, the fact that horses with bilateral palmar foot pain were not studied, the moderate camera resolution, and the failure to determine the results on different support surfaces. The operator was not blinded during the marking of the study areas. Time elapsed after perineural analgesia was used to evaluate the lameness score, which could have obscured problems with many more structures than the palmar heel region alone. Another limitation is that we did not use MRI, CT, or SCI to properly diagnose lesions of the navicular bone, DDFT, navicular bursa, or collateral and impaired ligaments in the studied and control groups.

5. Conclusions

The minimum temperature in the toe area presented an obvious increase in temperature following training sessions in horses with unilateral palmar foot pain, compared with the same area in the other non-lame limbs. The local temperature in the frog area for horses with palmar foot pain is decreased compared to the temperature observed in the non-lame limbs, both before and after training.

Based on the results obtained in this pilot study, we suggest a thermographic scanning of the toe and frog area to detect changes in the local temperature of the sole surface after training to discriminate the horses with palmar foot pain from non-lame ones.

Further investigations, on a larger number of horses, are required to clarify that the observed thermal pattern can be proposed as being characteristic for horses with palmar foot pain.

Supplementary Materials: The following supporting information can be downloaded at: https://www.mdpi.com/article/10.3390/biology12030423/s1, Figure S1: Lameness score; Figure S2: ROC curve frog area; Figure S3: ROC curve toe area; Tables S1–S6—Thermography results–Temperature for toe (El2) and frog (El1) areas.

Author Contributions: Conceptualization, methodology and study supervision, C.I.; data collection, manuscript preparation, design and writing—original draft preparation, C.Z.; data collection and investigation, R.D., L.S., P.N. and T.F.; statistical analysis and interpretation of data, C.R.; acquisition of data and manuscript review, B.S. and K.I. All authors have read and agreed to the published version of the manuscript.

Funding: This research paper is supported by the project "Increasing the impact of excellence research on the capacity for innovation and technology transfer within USAMVB Timisoara", code 6PFE, submitted in the competition Program 1—Development of the national system of research development, Subprogram 1.2—Institutional performance, Institutional development projects—Development projects of excellence in RDI.

Institutional Review Board Statement: The study was conducted according to the guidelines of the Declaration of Helsinki, and approved by the Bioethics Commission of Banat University of Agricultural Sciences and Veterinary Medicine "King Michael I of Romania" from Timisoara with No.51/07.06.2021.

Informed Consent Statement: Informed consent was obtained from the owners of the horses that were examined in the clinic and in the field.

Data Availability Statement: All the data obtained during the retrospective study are available in the Clinical Register of Large Animals from the Surgery Clinic.

Acknowledgments: The authors are grateful and thankful to the horse owners who participated in the study. All the horses owners give their consent to participate in the study.

Conflicts of Interest: The authors declare no conflict of interest.

References

1. Dabareiner, R.M.; Carter, G.K. Diagnosis, treatment, and farriery for horses with chronic heel pain. *Vet. Clin. N. Am. Equine Pract.* **2003**, *19*, 417–441. [CrossRef] [PubMed]
2. Dyson, S.; Marks, D. Foot pain and the elusive diagnosis. *Vet. Clin. N. Am. Equine Pract.* **2003**, *19*, 531–565. [CrossRef]
3. Blunden, A.; Dyson, S.; Murray, R.; Schramme, M. Histopathology in horses with chronic palmar foot pain and age-matched controls. Part 1: Navicular bone and related structures. *Equine Vet. J.* **2006**, *38*, 15–22. [CrossRef]
4. Dyson, S.; Murray, R.; Schramme, M.; Blunden, T. Current concepts of navicular bone. *Equine Vet. Educ.* **2011**, *23*, 7–39. [CrossRef]
5. Robinson, K.A.; Manning, S.T. Efficacy of a single –formula acupuncture treatment for horses with palmar heel pain. *Can. Vet. J.* **2015**, *56*, 1257–1260. [PubMed]
6. Coomer, R.; Thomas, H.; McKan, S. Current concepts of navicular syndrome: Diagnosis and treatment. *Practice* **2013**, *35*, 325–335. [CrossRef]
7. Turner, T.A. Diagnosis and treatment of the navicular syndrome in horses. *Vet. Clin. N. Am. Equine Pract.* **1989**, *5*, 131–144. [CrossRef]
8. Waguespack, R.; Hanson, R.R. Navicular syndrome in equine patients anatomy, causes, and diagnosis. *Compend. Contin. Educ. Vet.* **2010**, *32*, E7.
9. Matthew, S. Navicular disease ... remembrance of things past. *Vet. Comp. Orthop. Traumatol.* **2017**, *30*, V–VI. [CrossRef]
10. Turner, T.A. How to Subjectively and Objectively Examine the Equine Foot. *Proc. Am. Assoc. Equine Pract.* **2006**, *52*, 531–537.
11. Soroko, M.; Howell, K.; Zielinska, P. Application of thermography in racehorse performance. In Proceedings of the XV Congress of the European Association of Thermology, Wroclaw, Poland, 1–4 September 2021. [CrossRef]
12. Latif, Y.; Zafer, O.; Elif, D. Instrumentation of Thermography and its Applications in Horse. *J. Anim. Vet. Adv.* **2007**, *6*, 858–862.
13. Turner, T.A. Diagnostic Thermography. *Vet. Clin. N. Am. Equine Pract.* **2001**, *17*, 95–105. [CrossRef] [PubMed]

14. Jose, Q. *Application of Infrared Thermography in Sports Science*, 1st ed.; Springher: Cham, Switzerland, 2017; pp. 274–280. ISBN 978-3-319-47409-0. [CrossRef]
15. Fonseca, B.P.A.; Alves, A.L.G.; Nicoletti, J.L.M.; Thomassian, A.; Hussini, C.A.; Mikaik, S. Thermography and ultrasonography in back pain diagnosis of equine athletes. *J. Equine Vet. Sci.* **2006**, *11*, 507–516. [CrossRef]
16. Von Schweinitz, D.G. Thermographic diagnosis in equine back pain. *Vet. Clin. N. Am. Equine Pract.* **1999**, *15*, 161–177. [CrossRef]
17. Radaelli, V.; Bergero, D.; Zucca, E.; Ferrucii, F.; Costa, L.; Crosta, L.; Luzi, F. Use of thermography techniques in equines: Principles and applications. *J. Equine Vet. Sci.* **2014**, *34*, 345–350. [CrossRef]
18. Turner, T.A. Thermography as an aid to the clinical lameness evaluation. *Vet. Clin. N. Am. Equine Pract.* **1991**, *7*, 311–318. [CrossRef]
19. Widmer, W.R.; Buckwalter, K.A.; Fessler, J.F.; Hill, M.A.; Van Sickele, D.; Inavchevic, S. Use of radiography, computed tomoraphy and magnetic resonance imaging for evaluation of navicular syndrome in the horse. *Vet. Radiol. Ultrasound* **2000**, *41*, 108–116. [CrossRef]
20. Dyson, S. Radiological interpretation of the navicular bone. *Equine Vet. Educ.* **2011**, *23*, 73–87. [CrossRef]
21. Ross, M.W.; Dyson, S.J. *Diagnosis and Management of Lameness in the Horse*, 2nd ed.; Elsevier: St. Louis, MO, USA, 2010; pp. 78–79.
22. McGuigan, M.P.; Wilson, A.M. The effect of bilateral palmar digital nerve analgesia on the compressive force experienced by the navicular bone in horses with navicular disease. *Equine Vet. J.* **2001**, *33*, 166–171. [CrossRef]
23. Hoerdemann, M.; Smith, R.L.; Hosgood, G. Duration of action of mepivacaine and lidocaine in equine palmar digital perineural blocks in an experimental lameness model. *Vet. Surg.* **2017**, *46*, 986–993. [CrossRef]
24. Chanda, M.; Senarat, W.; Thongkam, E.; Kanthavichit, K.; Puangthong, C. Unilateral perineural anaesthesia on the lame leg facilitates the selection of z bar shoeing technique for the treatment of navicular syndrome. *J. Appl. Anim. Res.* **2019**, *47*, 154–158. [CrossRef]
25. Schumacher, J.; Schumacher, M.C.; Schumacher, J.; DeGraves, F.J. Diagnostic analgesia of the foot. *Equine Vet. Educ.* **2004**, *6*, 159–165.
26. Bidwell, L.A.; Brown, K.E.; Cordier, A.; Mullineaux, D.R.; Clayton, H.M. Mepivacaine local anesthetic duration in equine palmar digital nerve blocks. *Equine Vet. J.* **2010**, *36*, 723–726. [CrossRef] [PubMed]
27. Soroko, M.; Howell, K.; Dudek, K.; Wilk, I.; Zastrzezynska, M.; Janczarek, I. A pilot study into the utility of dynamic infrared thermography for measuring body surface temperature changes during treadmill exercise in horse. *J. Equine Vet. Sci.* **2018**, *62*, 44–46. [CrossRef]
28. Soroko, M.; Howell, K. Infrared thermography: Current applications in Equine Medicine. *J. Equine Vet. Sci.* **2018**, *60*, 90–96. [CrossRef]
29. Soroko, M.; Davies, M. *Equine Thermography in Practice*; Replika Press PVT Ltd.: Sonipat, India, 2016; pp. 14–75. ISBN 9781780647876. [CrossRef]
30. Ciutacu, O.; Tanase, A.; Miclaus, I. Digital Infrared thermography in assessing soft tissues injuries on sport equines. *USAMV-CN* **2006**, *63*, 228–233.
31. Agne, B. Diagnosis and treatment of foot infections. *J. Equine Vet. Sci.* **2010**, *30*, 510–512. [CrossRef]
32. Buchner, H.H.F.; Savelberg, H.H.C.M.; Schamhardt, H.C.; Barneveld, A. Limb movement adaptations in horses with experimentally induced fore- or hindlimb lameness. *Equine Vet. J.* **1996**, *28*, 63–70. [CrossRef]
33. Merkens, H.W.; Schamhardt, H.C. Evaluation of equine locomotion during different degrees of experimentally induced lameness I: Lameness model and quantification of ground reaction force patterns of the limbs. *Equine Vet. J.* **1988**, *20*, 99–106. [CrossRef]
34. Soroko, M.; Henklewski, R.; Filipowski, H.; Jodkowska, E. The effectiveness of Thermographic Analysis in Equine Orthopedics. *J. Equine Vet. Sci.* **2013**, *33*, 760–762. [CrossRef]
35. Weishaupt, M.A.; Weistner, T.; Herman, P.H.; Jordan, P.; Auer, A. Compensatory load redistribution of horses with induced weght-bearing forelimb lameness trotting on a treadmill. *Vet. J.* **2006**, *171*, 135–146. [CrossRef] [PubMed]
36. Eddy, A.L.; van Hoogmoed, L.M.; Snyder, J.R. The role of thermography in the management of equine lameness. *Vet. J.* **2001**, *162*, 172–181. [CrossRef] [PubMed]
37. Holmes, L.C.; Gaughan, E.M.; Gorondy, D.A.; Hogge, S.; Spire, M.F. The effect of perineural anesthesia on infrared thermographic images of the forelimb digits of normal horses. *Can. Vet. J.* **2003**, *44*, 392–396. [PubMed]
38. Simon, E.L.; Gaughan, E.M.; Epp, T.; Spire, M. Influence of exercise on thermographically determined surface temperatures of thoracic and pelvic limbs in horses. *J. Am. Vet. Med. Assoc.* **2006**, *229*, 1940–1944. [CrossRef]
39. Talukdar, A.H.; Calhoun, M.L.; Stinson, A.W. Sweat glands of the horse: A histologie study. *Am. J. Vet. Res.* **1970**, *31*, 2179–2190.

biology

MDPI

Article

SPIM-Flow: An Integrated Light Sheet and Microfluidics Platform for Hydrodynamic Studies of *Hydra*

Per Niklas Hedde [1,*], Brian T. Le [2], Erika L. Gomez [2], Leora Duong [3], Robert E. Steele [4] and Siavash Ahrar [2,5,*]

1 Beckman Laser Institute and Medical Clinic, University of California Irvine, Irvine, CA 92612, USA
2 Department of Biomedical Engineering, CSU Long Beach, Long Beach, CA 90840, USA
3 Department of Molecular Biology and Biochemistry, University of California Irvine, Irvine, CA 92697, USA
4 Department of Biological Chemistry, University of California Irvine, Irvine, CA 92697, USA
5 Department of Physics and Astronomy, University of California Irvine, Irvine, CA 92697, USA
* Correspondence: phedde@uci.edu (P.N.H.); siavash.ahrar@csulb.edu (S.A.)

Simple Summary: This work demonstrates an inexpensive light sheet microscope that can be readily integrated with modern microfluidics. The integration between these two technologies is critical for investigating the hydrodynamics of aquatic organisms. As a case study, we examined the hydrodynamics of *Hydra* in millimeter-sized chambers. The platform was used to study *Hydra*'s response to flow and detachment from channel surfaces. Broader access to light sheet microscopy, particularly integrated with fluidics, could be critical for providing exciting opportunities to investigate the hydrodynamics of freely moving model organisms.

Abstract: Selective plane illumination microscopy (SPIM), or light sheet microscopy, is a powerful imaging approach. However, access to and interfacing microscopes with microfluidics have remained challenging. Complex interfacing with microfluidics has limited the SPIM's utility for studying the hydrodynamics of freely moving multicellular organisms. We developed SPIM-Flow, an inexpensive light sheet platform that enables easy integration with microfluidics. We used SPIM-Flow to investigate the hydrodynamics of a freely moving *Hydra* polyp via particle tracking in millimeter-sized chambers. Initial experiments across multiple animals, feeding on a chip (*Artemia franciscana* nauplii used as food), and baseline behaviors (tentacle swaying, elongation, and bending) indicated the organisms' health inside the system. Fluidics were used to investigate *Hydra*'s response to flow. The results suggested that the animals responded to an established flow by bending and swaying their tentacles in the flow direction. Finally, using SPIM-Flow in a proof-of-concept experiment, the shear stress required to detach an animal from a surface was demonstrated. Our results demonstrated SPIM-Flow's utility for investigating the hydrodynamics of freely moving animals.

Keywords: light sheet; microfluidics; *Hydra*; hydrodynamics

1. Introduction

Need for SPIM: Selective plane illumination microscopy (SPIM), or light sheet microscopy, is a powerful approach with high spatio-temporal resolution for biological samples imaging [1]. Typically, the excitation arm (i.e., the light sheet path) of a microscope is arranged at a 90-degree angle relative to the detection arm (i.e., the camera path) [2]. In an observation plane, an excitation light is restricted to a thin illumination plane via cylindrical optics or other beam shaping methods. This one-dimensional confinement provides true optical sectioning, and it limits the phototoxicity and photobleaching of the sample, thus enabling long-term imaging [1]. Many SPIM platforms have been demonstrated [3]. Examples include open-source approaches [4,5], systems that use a single objective [6], and even systems that use mobile phones for imaging [7]. SPIM has enabled biological discoveries across sub-cellular imaging [8,9], developmental studies (across vertebrates, invertebrates, and plant model systems) [10–12], and optically cleared tissues (e.g., the entire brains of small animals) [13]. While SPIM provides a powerful approach for long-term imaging, samples are often housed in static conditions. Significant modifications to fluidics are required when fluidics are used to accommodate light sheet microscopy. Thus, interfacing SPIM with conventional microfluidics could further enhance the approach's utility by allowing the dynamic manipulation of a sample's microenvironment and broadening the use of the technique [14].

Integrating SPIM with fluidics: Various strategies have been explored to integrate fluidics with light sheet microscopy [15,16]. Such integration has enabled cytometry platforms [17–20], the generation and visualization of droplets [21], and visualizing *C. elegans* and *Drosophila* embryos [22,23]. For example, a recent and exciting study by Vanwalleghem et al. demonstrated the strength of integration for studying larval zebrafish brain-wide activity while using flow as a stimulus [24]. In previous work, we explored using a light sheet configuration (based on an inverted epifluorescence microscope) which could accommodate chambers with flow [25]. Despite this progress, integrating SPIM with microfluidics has remained difficult. For example, the integration requires modifications to the microscope or significant modifications to the microchambers [15]. Additionally, the total cost (for example, for approaches that augment high-end research microscopes with light sheet capabilities) and the complexity of establishing light sheet systems have precluded their wide adoption across the microfluidic and lab-on-a-chip community. In particular, light sheet illumination can be of great value for microfluidic applications which require particle image tracking and velocimetry. To address this gap, we developed SPIM-Flow, a simple and inexpensive system that readily integrates light sheet microscopy with fluidics. As a proof-of-concept study, we used SPIM-Flow to investigate the hydrodynamics of a freely moving *Hydra* polyps in a millimeter-sized chamber (4 mm wide and 1.6 mm in height). Specifically, instead of utilizing the 3D imaging capabilities of the approach, SPIM-Flow was used to side-illuminate the freely moving organisms, and micron-sized tracer particles were used to visualize the hydrodynamics.

Hydra and microfluidics—an approach to investigate hydrodynamics: *Hydra* is a freshwater cnidarian known for its remarkable regenerative abilities [26]. Its transparent, tube-shaped body is divided into three regions: head, body column, and foot. The head includes the tentacles and hypostome—a dome-like structure containing the mouth opening at its apex. *Hydra* uses its foot (i.e., a basal disk) to attach to surfaces. The shape and movement of the body are controlled via a hydrostatic skeleton where fluid pressure transmits forces [26,27]. With these characteristics, *Hydra* is an excellent system to use to investigate biomechanics and hydrodynamics. A recent study, for example, using imaging and machine learning, showed that the behavioral repertoire of *Hydra* could be divided into six components (i.e., elongation, tentacle sway, body sway, bending, contraction, and somersaulting) [28]. In other studies, it was shown that *Hydra* must tear a hole through its epithelial tissue to open its mouth [29,30]. However, these investigations were conducted under static conditions. Using fluid chambers would allow one to expand upon the current studies by modulating the microenvironments' physical (e.g., flow) or chemical (e.g., transitory drug delivery) compositions. Additionally, previous applications of light sheet microscopy to study *Hydra* have used static chambers [31,32].

In another recent study, Badhiwala et al. developed three microfluidic systems to study *Hydra* [33]. These included a chamber to constrain the body column and enable electrophysiology, a perfusion chamber for constrained locomotion, and a quasi-2D plane behavioral chamber (200–600 μm in height). More recently, the investigators used a double-layer microfluidic system to mechanically stimulate *Hydra* (via pneumatic valves) and measure neuronal responses via calcium imaging [34]. Microfluidic platforms enabled neuronal studies that required animal immobilization or small chambers. However, these restrictions limit an organism's movement and may negatively impact its health. In our study, by taking advantage of SPIM-Flow, we sought to investigate *Hydra*'s response to flow without dramatically restricting its movement. Chambers with millimeter-sized dimensions (4 mm wide and 1.6 mm tall) were constructed. Next, we utilized the integration of microfluidics to modulate the hydrodynamic environment, for example, by introducing flow or prey. We used the light sheet system to visualize the hydrodynamics of the animal's movement. Our studies suggested that *Hydra* typically remained in an elongated or swaying state inside the chambers without flow. *Hydra* responds to flow initiation via contraction or tentacle swaying. Moreover, *Hydra* typically reorients by body bending (head and tentacles) in the flow direction. Finally, utilizing the system, the shear stress (at the surface of the chamber) required to detach *Hydra* from devices was measured.

2. Materials and Methods

Optical System: The optical system of SPIM-Flow is composed of excitation and detection arms (Figure 1). To create the light sheet, similar to our prior efforts [8], the central part of the elliptical beam of a laser diode (wavelength 488 nm and 30 mW, variant dot/point and TTL output, LambdaWave Turquoise Laser) was cut with an iris (SM1D12, Thorlabs, Newton, NJ, USA). Then, a cylindrical lens (f = 12.7 mm, LJ1942L1-A, Thorlabs) was used to focus and form a sheet of light. These components were housed inside a custom-designed, 3D-printed enclosure. A cannon-like design (i.e., a light-cannon) was developed to bring the cylindrical lens close to the fluidic chip. The 3D-printed enclosure containing the optical components for excitation was then mounted on an XYZ micromanipulator. A light sheet of 7 μm thickness was created utilizing these components (see Supplementary Information for analysis). The excitation arm was mounted on an optical breadboard at a 90-degree angle relative to the detection arm. The detection arm included a custom-designed 3D-printed chip holder to bring the fluidic chip close to the light-cannon. Fluorescent signals were collected perpendicular to the excitation with an objective lens (10×, NA 0.25, Newport Spectra-Physics, Milpitas, CA, USA), followed by an emission filter (500 nm long pass, ET500lp, Chroma, Bellows Falls, VT, USA) and a tube lens (f = 50 mm, AC254-050-A, Thorlabs) for imaging with a CMOS camera (20 MP, FLIR Blackfly S-U3-200S6C-C, Teledyne FLIR, Goleta, CA, USA). A SPIM-Flow system-level diagram is provided (Figure 1A). The system was housed on a small optical breadboard (300 mm × 300 mm) and was inside an opaque enclosure to shield the setup from external light. Micro-Manager software [35] was used to control the camera. Designs for the 3D printed components are available from a corresponding OSF page. Unlike systems that augment a standard microscope (e.g., using an inverted wide-file standard microscope as the detection system) [36], SPIM-Flow is a standalone and portable instrument.

Microfluidic systems: A desktop CO_2 laser cutter was used to build the millimeter-sized molds used in the study instead of using the conventional lithographic approaches. This choice was informed by the size of *Hydra* (an average length of 10 mm) and our goal to investigate their unhindered hydrodynamics. To this aim, channels were designed to be 4 mm wide and 5 cm long. Acrylic sheets (1/16th inch, 1.6 mm thickness) were laser-cut (40 W Glowforge desktop systems). Plastic parts were retrieved from the sheet and permanently attached to a flat base (either a plastic part or a glass slide) via an adhesive. The attachment prevented the unwanted PDMS accumulation under the mold during the silicone casting. After the mold fabrication, conventional protocols were used to cast silicone parts from the mold (1:10 linker to the base mass ratio, followed by baking the molds on hotplates at 85 °C until the PDMS was fully cured). Individual PDMS devices were cut from the mold using craft knives, and inlet/outlet holes were introduced using disposable (1 or 2 mm in diameter) biopsy punches.

Next, each PDMS part was plasma-bonded to a microscope glass slide via a Harrick Basic Plasma cleaner. The PDMS parts were placed on the edge of the glass slide (instead of at the center) to minimize the distance between the channel and the light cannon (Figure 1B,C). Additionally, steps were taken to remove any roughness from the outer edges of the devices to minimize potential imaging artifacts (some roughness can be introduced to the outer edges of PDMS parts while cutting them via a craft knife). In a typical epifluorescent illumination, such roughness is of no concern since the path of light is through the device top (typically, the PDMS) and glass bottom. In our system, however, the chamber was illuminated through a PDMS wall and viewed from the device's top. Additional uncured PDMS was applied to the device's outer wall using a craft knife after plasma-bonding to remove any roughness. Next, the devices, with the side with fresh PDMS facing up, were left at room temperature for 3–4 h. This pause allowed the PDMS to spread evenly and partially cure. Devices were then backed at 95 °C on hotplates to ensure that the added PDMS was entirely cured. During the early experiments, we also attached coverslips to the device roofs. However, later experiments demonstrated that the coverslip did not significantly improve imaging.

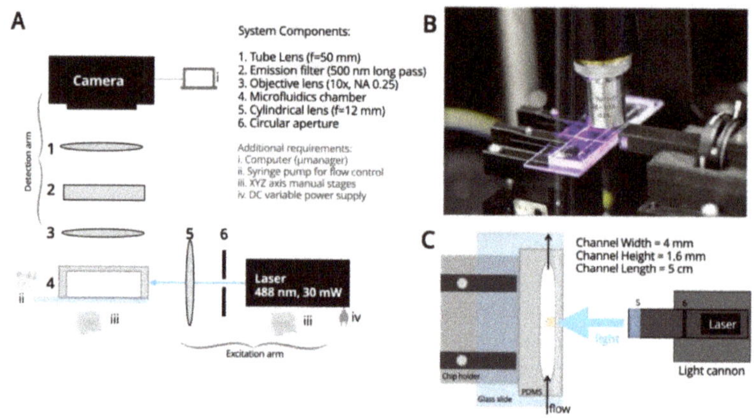

Figure 1. SPIM-Flow system diagram. (**A**) Simple and inexpensive SPIM designed for compatibility with fluidics. (**B**) A photograph of the excitation arm (i.e., the light cannon), microfluidic chamber held on the 3D printed stage, and the objective lens collecting fluorescent signals from the chamber. (**C**) Diagram of the PDMS chamber with its critical dimensions, the stage, and the light cannon. Steps were taken to remove any roughness from the side of the chamber that accepted the light and to remove any rough edges that could create imaging artifacts.

Hydra culture: All experiments were carried out using the PT1 transgenic line of *Hydra* vulgaris [37]. PT1 contains two transgenes. One of the transgenes expresses green fluorescent protein under the control of the promoter from the gene encoding the Hym176B neuropeptide. The other transgene expresses the gene for the red fluorescent protein, the DsRed2 gene, under the control of an actin gene promoter. This line expresses green fluorescent neurons and red fluorescent epithelial cells. PT1 was maintained in *Hydra* medium 4.0, which was prepared using house deionized water and that contained 1 mM calcium chloride, 0.33 mM magnesium sulfate, 0.5 mM sodium bicarbonate, and 0.03 mM potassium chloride. The animals were fed once a week with the nauplii of *Artemia franciscana* from San Francisco Bay (Brine Shrimp Direct, Ogden, UT, USA). The *Hydra* cultures were kept in an incubator at 18 °C on a 12 h light/12 h dark cycle.

Loading the *Hydra* in microfluidics: A simple protocol was developed to minimize potential damage and increase the successful loading of the animals. First, a device was half-filled with fresh medium. A glass Pasteur pipette was used to retrieve an animal from a culture tube. The biggest challenge was ensuring *Hydra* adhered to the glass transfer pipette, since *Hydra* can rapidly adhere and remain firmly attached to surfaces. Therefore, it was essential to transfer each *Hydra* quickly. The animals were placed directly inside a device inlet or on top of an inlet with excess medium from the pipette. Next, the liquid was withdrawn with a micropipette (1 mL) or a syringe to draw the animal with the additional liquid into the chamber. Each *Hydra* was positioned toward the chamber's center for ease of imaging. Red fluorescent beads (1 μm diameter, Fluoro-Max Polymer microspheres,

ThermoFisher Scientific, Waltham, MA, USA) were added to the medium to enable flow and hydrodynamic visualizations. After establishing the protocol, the animal loading process had a near-perfect success rate.

After loading an animal, the chamber was placed and secured on the 3D-printed chip holder. *Hydra* medium with beads was delivered to the chamber using an external syringe pump (KDS Legato, 210P series) via 3, 10, or 20 mL syringes. Typical of all microfluidics experiments, care was taken to ensure there were no air bubbles upstream of the device. However, small bubbles did not lead to animal death. For example, early in the studies, air bubbles were accidentally introduced into two chambers. In each case, the animal rapidly responded to the air bubble and the pressure by contracting/bending. The animals were able to survive these exposures.

3. Results

3.1. Health and Feeding

At the start of the study, to verify the animals' health in the SPIM-Flow, feeding behaviors were investigated. After loading a *Hydra* onto a chip, multiple prey animals (*Artemia franciscana*) were added to the chamber. First, imaging without the light sheet capabilities was used to investigate the potentially harmful effects of the chamber. Each *Hydra* could successfully capture and eat multiple (two to three) *Artemia nauplii* inside the chamber. Next, feeding behavior was observed while using the SPIM imaging capabilities across multiple independent animals (Supplementary Figure S1 and Supplementary Video S1, though the video and images are from two different animals). After feeding, the *Hydra* ignored the remaining prey in the chamber. The feeding results (across N = 4 different animals) suggested that the fluid chambers and the light sheet did not negatively impact the organisms' health. Subsequent hydrodynamic experiments typically took 4 to 5 h. However, it was confirmed that the *Hydra* could be kept alive on the same chip for multiple days.

3.2. Static Conditions

After examining the animals' health, we sought to investigate the behavioral repertoire of the *Hydra* without flow. The animals' movements were unhindered, given the millimeter-sized dimensions of the fluidic chambers. However, some animals could stretch across the entire 4 mm-width of the chamber. Utilizing the XYZ-stages, light sheet illumination could be positioned based on an animal's positioning. Previous studies have established that *Hydra* respond to light as a stimulus [38–41]. Therefore, the experiments were started 5 min after the introduction of the light sheet. Additionally, the light was kept on during most of the recording session. SPIM's capability to limit the light exposure (here, only to the 7 μm-thick illuminated sheet) prevented the potential of photobleaching or phototoxicity.

Most animals adhered to the surfaces (PDMS walls) closer to the direction of the excitation light (wavelength of 488 nm). Without flow, behaviors ranging from elongation to tentacle movement/swaying were observed (Figures 2 and 3, Supplementary Figure S2, and Supplementary Video S2). Most animals would remain in an elongated or tentacle-swaying state. These observations further supported the assumptions regarding the animals' overall health and helped demonstrate that various repertoires of the animals' movements could be captured via the SPIM-Flow.

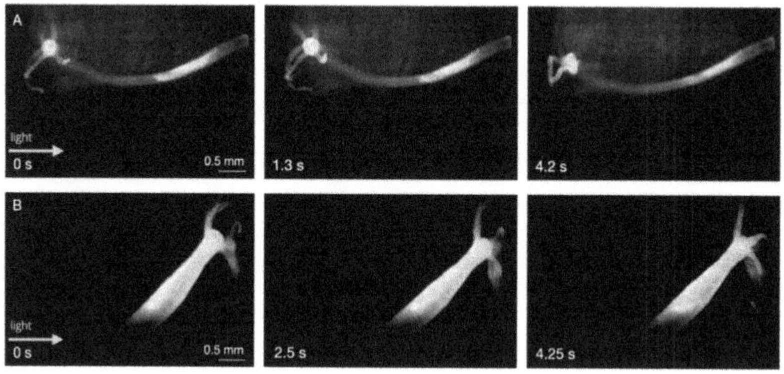

Figure 2. Baseline behavior during static conditions as an indicator of overall health. (**A**) Elongation and tentacle movement. The head (hypostome and tentacle ring) are also visible. (**B**) A different animal in an elongated state inside a chamber. Using SPIM-Flow, GFP-positive neurons could be visualized, depending on the orientation and state of the *Hydra* (excitation 488 nm and emission 500 nm long pass).

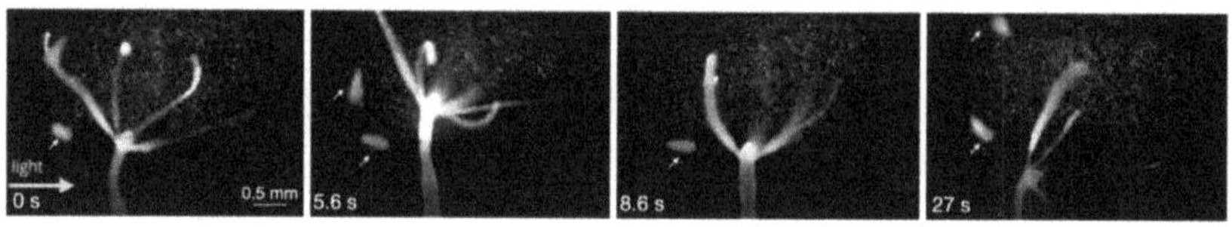

Figure 3. Baseline behaviors (bending and tentacle movement) during static conditions. Yellow arrows point to *Artemia franciscana* nauplii that were ignored by the *Hydra* after capturing 3 nauplii earlier.

3.3. Response to Flow

Next, taking advantage of microfluidics, *Hydra*'s response to flow was investigated. *Hydra* live in bodies of freshwater that may experience local flow due to environmental perturbations or currents (in streams and rivers). Therefore, tools to investigate their response to flow could provide valuable insights into the organisms' biomechanical and hydrodynamic lifestyles. In their study, Badhiwala et al. [33] commented that these animals typically bend in the flow direction. We sought to investigate this observation by utilizing our larger chambers (1.6 mm in height and 4 mm wide). Our results (across four independent animals) suggested that *Hydra* typically respond to flow (2 mL/h) by bending and swaying their tentacles in the flow direction (tentacle-swaying in the direction of the flow (Supplementary Video S3). Reversing the direction of flow also resulted in *Hydra* redirecting accordingly (Figure 4). The results also suggested that the animals could immediately respond to the flow initiation via rapid contraction or tentacle-swaying in the direction of the flow (Supplementary Figure S3).

Next, two previously described image analysis approaches were used to examine the animals' responses to flow closely. First, to track the movement of the animals, a markerless pose estimation software (i.e., DeepLabCut2.0) was used [42]. To this aim, locations on the body and the tentacles were identified across video recordings. Utilizing DeepLabCut, the movements of these positions were tracked to generate 2D skeletons of the animals (tentacle-swaying in the direction of the flow (Supplementary Video S4) and plots detailing the changes in the positions of previously identified locations (Figure 5). In the tracking plots, squares represented the starting position, and the triangle represented the final positions for a given body part. The color map indicates the movement of the marker through time (as demonstrated from the first and last frame of the video). Pose estimation software provided opportunities to analyze the results from the SPIM-Flow system experiments. Next, a previously described flow visualization algorithm (i.e., FlowTrace) was used to better visualize the flow features (primarily due to external flow) from the SPIM-Flow recordings. FlowTrace is a simple algorithm that enables the extraction of flow features from videos [43,44]. Unlike typical pipelines for particle image velocimetry, FlowTrace is not computationally taxing. Here, the algorithm was used a FIJI plugin. Therefore, a rapid analysis could provide critical hydrodynamic insights from the SPIM-Flow recordings. The results for these visualizations are provided in Supplementary Figure S4 and Video S5. The videos include both the original recording and the pathline visualizations.

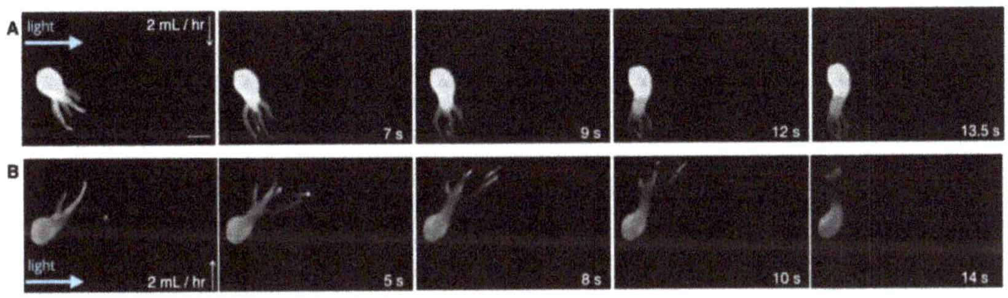

Figure 4. Response to flow. Our results suggested that *Hydra* typically bend and sway their tentacles in the direction of the flow (head aways from the direction of the flow). (**A**) Initiation of the flow and redirection of the animal towards the flow. The white arrow demonstrates the flow direction. The scale bar is 0.5 mm. (**B**) Reversing the flow direction led to the animal redirecting with the new flow direction.

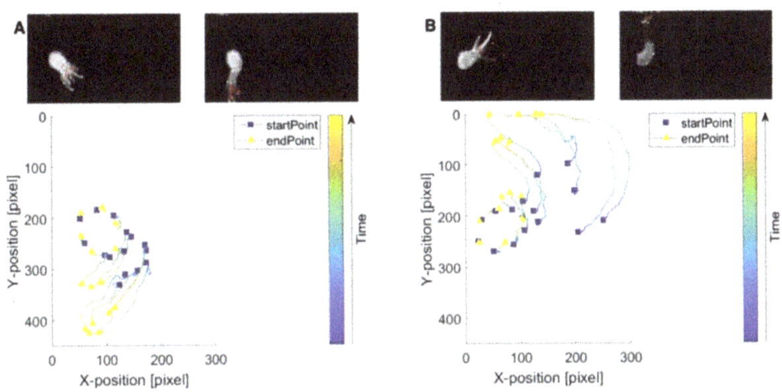

Figure 5. Tracking *Hydra*'s movements in response to flow and pseudo-skeleton visualizations. (**A**) Tracking *Hydra*'s responses to the initiation of flow. Key anatomical features were identified, a pseudo-skeleton was generated, and the movement of features was tracked over space (here, the units are pixels). (**B**) Response to reversing the direction of flow. In each case, tracking and pseudo-skeletons were generated via DeepLabCut. The integration of SPIM-Flow with pose estimation analysis provided pathways for biomechanics investigations. The starting position is highlighted with a blue square. The final position is highlighted with a yellow triangle. The colors indicate the position in time.

3.4. Adhesion and Detachment from Surfaces

Finally, using SPIM-Flow, surface shear stress required to detach a *Hydra* specimen from a glass substrate were measured. Investigating *Hydra*'s ability to adhere to surfaces under aqueous environments is of interest for both biomechanical and hydrodynamics. These organisms use adhesive composites (glue) to adhere to various surfaces [45]. Flow chambers could provide a valuable approach to quantitatively measure the shear stress and visualize the flow profiles required to detach *Hydra*. A previous study [46] used a large laminar-flow chamber (tanks) to estimate the forces required to detach these animals from various surfaces. The results suggested that the animal's size and the substrate play roles in the force required to detach an animal from a surface. However, tracking the fluid elements via particle tracers was absent from the previous system. In this study, we sought to demonstrate SPIM-Flow's utility for investigating adhesion/detachment. Our platform could provide fine control over the microenvironment and visualize tracer particles, as well as the animals. A previously fed *Hydra* was used to investigate the animal's detachment from a glass surface. The volumetric flow rate (Q) was progressively increased until the *Hydra* specimen was detached from a surface. The animal was rested for 5 min between testing each condition (10 mL/h, 20 mL/h, and 50 mL/h). In our experiments, the animal detached at 50 mL/h. Similar to previously developed analysis [47], the channel's shear stress at the glass surface (bottom of the chamber) was calculated via:

$$\sigma_S = \frac{6Q\mu}{wh^2}, \quad (1)$$

where Q is the volumetric flow rate (mL/h), μ is the viscosity of *Hydra*'s medium (estimated as the viscosity of the water at 1.002 mPa.s at 20 degrees Celsius), w is the channel width (4 mm), and h is the channel height (1.6 mm). Given the applied low rates of 10–50 mL/h, the corresponding surface shear stress ranged from approximately 1–10 mPa (1.63–8.15 mPa).

4. Discussion

Access to light sheet imaging and interfacing the conventional systems with fluidics have remained challenging. We developed an inexpensive light sheet microscope designed for ready integration with microfluidics (SPIM-Flow) to address these challenges. SPIM-Flow was made by integrating off-the-shelf optics and 3D-printed enclosures. Therefore, the system required no additional optical instruments for its operation.

This study used SPIM-Flow to investigate the hydrodynamics of freely moving *Hydra* polyps. In particular, the system was used as a miniaturized particle imaging/tracking instead of utilizing the 3D sectioning of the SPIM. To this aim, light sheet imaging enabled an easy approach for the visualization of 1 μm tracer beads. In comparison, the animals were also imaged inside the same microfluidic systems via a standard fluorescent microscope (Nikon Eclipse Ti-E microscope 10X objective—Supplementary Figure S5). Visualizing the beads and the animal (due to the high signal from the animal and the background volume) was challenging with a standard microscope. Additionally, the standard system would illuminate the entire chamber volume throughout the recording session, which could lead to phototoxicity. Our experiments (feeding and baseline behaviors) supported the assumption that the platform did not impact the animals' health. Feeding on a chip suggested the possibility for the long-term culture and investigation of *Hydra* inside the flow chambers. Next, microfluidics was used to investigate *Hydra's* response to flow. The results across multiple animals (N = 4) suggested *Hydra* responds to flow by bending in the flow direction (i.e., an animal's head is facing away from the direction and its tentacles are elongated in the direction of the flow (see Figure 4). Reversing the direction of flow led to animals' reorientation. Next, SPIM-Flow's compatibility with two image analysis approaches was demonstrated. In the first analysis, motion tracking software (DeepLabCut) was used to track the location of anatomical landmarks and generate pseudo-skeletons in response to flow. In future studies, landmark tracking could be used to estimate displacements and the speed of movements. Pseudo-skeletons could enable biomechanical studies related to quantifying contractile behaviors. Moreover, DeepLabCut supports automated image annotations with limited user input. In the second image analysis approach, a visualization software (FlowTrace) was used to better demonstrate the tracer beads' movement, mainly due to the external flow. This study was particularly aimed at investigating the role of flow as environmental stimuli. The combination of SPIM-Flow and FlowTrace in future studies (using higher magnification and larger beads) could enable the investigation of flow patterns generated by *Hydra* alone, without external flows. Finally, in a proof-of-concept experiment, SPIM-Flow was used to visualize and measure the shear forces (50 mL/h, corresponding to surface shear stress of approximately σ_s = 10 mPa) required to detach a *Hydra* specimen from the glass surface of the chamber. Unlike the previous study [46], which used a large laminar-flow tank, SPIM-flow provided a more accessible approach for the fine control and hydrodynamic analysis of bioadhesion.

SPIM-Flow provided an easy approach for particle tracking/imaging. However, as was apparent in some recordings (see Figure 4), the *Hydra* could obstruct the illumination—due to the side illumination—by casting shadows, creating dark regions in certain portions of the chamber. In future studies, this issue could be addressed by repositioning the light cannon (i.e., illuminating from the opposite side after *Hydra*'s attachment) or, alternatively, by illuminating from both sides. It is also important to acknowledge that the 3D structure of *Hydra* and its fast movements make it difficult to keep the entire animal in focus. This task can be accomplished by restricting the animals, similar to the microfluidic systems previously developed by Badhiwala et al. [33]. However, unlike prior microfluidics with chambers of 200–600 µm in height, the chambers presented in this study did not restrict the movement of the animals (1.6 mm in height).

5. Conclusions

In the future, we plan to use SPIM-Flow to further investigate the hydrodynamics of aquatic organisms (including *Hydra*) by quantifying displacements generated by the animals and systematically exploring the animals' responses to pulsatile flow. Our preliminary results suggested that pulsatile flow also resulted in an animal stretching toward the flow's direction. Additionally, light (focused as a sheet) could be used as a stimulus in the systems. In this context, SPIM-Flow could serve as a valuable approach for integrating various environmental stimuli (light, forces/flow, and chemical factors). Microfluidics combined with conventional imaging has provided an excellent platform for studying model systems such as zebrafish, *Drosophila embryos*, and *C. elegans* [48,49]. Other emerging model systems could (and, indeed, have started to) benefit from the needs-driven systematic integration of micro- and milli-fluidics with novel imaging techniques [50–52]. Moreover, SPIM-Flow, as a simple and inexpensive platform, can contribute to these studies focusing on the hydrodynamics of freely moving aquatic organisms.

Supplementary Materials: The following supporting information can be downloaded at: https://www.mdpi.com/article/10.3390/biology12010116/s1. Supplementary section—analysis of light sheet properties. List of videos and brief descriptions. Supplementary figures. Figure S1: *Hydra* feeding behavior as an indicator of health; Figure S2: *Hydra* baseline behavior in static conditions as an indicator of overall health; Figure S3: Responses to the initiation of flow; Figure S4: Hydrodynamic visualization via FlowTrace; Figure S5: Imaging *Hydra* via a fluorescent microscope; Video S1: *Hydra* feeding behavior (*Artemia franciscana nauplius*, or brine shrimp, used as food); Video S2: *Hydra* tentacle-sway behavior; Video S3: *Hydra* response to flow; Video S4: *Hydra* motion tracking and pseudo-skeleton; Video S5: Hydrodynamic pathline visualizations and sample visualizations (original video and FlowTrace).

Author Contributions: Conceptualization, S.A. and P.N.H.; methodology, S.A., R.E.S. and P.N.H.; software, S.A. and B.T.L.; validation, S.A., P.N.H. and B.T.L.; formal analysis, S.A., P.N.H. and B.T.L.; investigation, S.A., P.N.H., R.E.S., E.L.G. and L.D.; resources, S.A., P.N.H. and R.E.S.; data curation, S.A. and P.N.H.; writing—original draft preparation, S.A. and P.N.H.; writing—review and editing, all authors; visualization, S.A. and P.N.H.; supervision, S.A.; project administration, S.A.; funding acquisition, S.A. and P.N.H. All authors have read and agreed to the published version of the manuscript.

Funding: This research was funded by CSULB startup funds and a CSUPERB New Investigator grant to SA. Additionally, this work was supported by the National Institutes of General Medical Sciences R21GM135493 to PNH. LD was supported by an NSF GRFP (base number 2021317020). The content is solely the responsibility of the authors and does not necessarily represent the official views of the funding agencies.

Institutional Review Board Statement: Not applicable.

Informed Consent Statement: Not applicable.

Data Availability Statement: Data is available upon request.

Acknowledgments: The authors express their sincere gratitude to Albert Siryaporn for his support throughout this study. The authors express their gratitude to Tim Smith for his support and helpful discussions regarding the DeepLabCut 2.0 analysis and computation approaches.

Conflicts of Interest: The authors declare no conflict of interest.

References

1. Stelzer, E.H.K.; Strobl, F.; Chang, B.-J.; Preusser, F.; Preibisch, S.; McDole, K.; Fiol, R. Light sheet fluorescence microscopy. *Nat. Rev. Methods Prim.* **2021**, *1*, 73. [CrossRef]
2. Huisken, J.; Swoger, J.; Del Bene, F.; Wittbrodt, J.; Stelzer, E.H.K. Optical sectioning deep inside live embryos by selective plane illumination microscopy. *Science* **2004**, *305*, 1007–1009. [CrossRef] [PubMed]
3. Olarte, O.E.; Andilla, J.; Gualda, E.J.; Loza-Alvarez, P. Light-sheet microscopy: A tutorial. *Adv. Opt. Photonics* **2018**, *10*, 111–179. [CrossRef]
4. Voigt, F.F.; Kirschenbaum, D.; Platonova, E.; Campbell, R.A.A.; Kastli, R.; Schaettin, M.; Egolf, L.; Van der Bourg, A.; Bethge, P.; Haenraets, K.; et al. The mesospim initiative: Open-source light-sheet microscopes for imaging cleared tissue. *Nat. Methods* **2019**, *16*, 1105–1108. [CrossRef] [PubMed]
5. Gualda, E.J.; Pereira, H.; Vale, T.; Estrada, M.F.; Brito, C.; Moreno, N. Spim-fluid: Open source light-sheet based platform for high-throughput imaging. *Biomed. Opt. Express* **2015**, *6*, 4447–4456. [CrossRef]
6. Yang, B.; Lange, M.; Millett-Sikking, A.; Zhao, X.; Bragantini, J.; VijayKumar, S.; Kamb, M.; Gómez-Sjöberg, R.; Solak, A.C.; Wang, W.; et al. Daxi—High-resolution, large imaging volume and multi-view single-objective light-sheet microscopy. *Nat. Methods* **2022**, *19*, 461–469. [CrossRef]
7. Hedde, P.N. minispim—A miniaturized light-sheet microscope. *ACS Sens.* **2021**, *6*, 2654–2663. [CrossRef]
8. Hedde, P.N.; Stakic, M.; Gratton, E. Rapid measurement of molecular transport and interaction inside living cells using single plane illumination. *Sci. Rep.* **2014**, *4*, 7048. [CrossRef]
9. Sapoznik, E.; Chang, B.-J.; Huh, J.; Ju, R.J.; Azarova, E.V.; Pohlkamp, T.; Welf, E.S.; Broadbent, D.; Carisey, A.F.; Stehbens, S.J.; et al. A versatile oblique plane microscope for large-scale and high-resolution imaging of subcellular dynamics. *Elife* **2020**, *9*, e57681. [CrossRef]
10. Lemon, W.C.; Pulver, S.R.; Höckendorf, B.; McDole, K.; Branson, K.; Freeman, J.; Keller, P.J. Whole-central nervous system functional imaging in larval drosophila. *Nat. Commun.* **2015**, *6*, 7924. [CrossRef]
11. Silvestri, L.; Bria, A.; Costantini, I.; Sacconi, L.; Peng, H.; Iannello, G.; Pavone, F.S. Micron-scale resolution optical tomography of entire mouse brains with confocal light sheet microscopy. *J. Vis. Exp.* **2013**, *80*, e50696. [CrossRef]
12. Von Wangenheim, D.; Hauschild, R.; Friml, J. Light sheet fluorescence microscopy of plant roots growing on the surface of a gel. *J. Vis. Exp.* **2017**, *119*, e55044. [CrossRef]
13. Ueda, H.R.; Dodt, H.-U.; Osten, P.; Economo, M.; Chandrashekar, J.; Keller, P. Whole-brain profiling of cells and circuits in mammals by tissue clearing and light-sheet microscopy. *Neuron* **2020**, *106*, 369–387. [CrossRef]
14. Zhou, P.; He, H.; Ma, H.; Wang, S.; Hu, S. A review of optical imaging technologies for microfluidics. *Micromachines* **2022**, *13*, 274. [CrossRef]
15. Albert-Smet, I.; Marcos-Vidal, A.; Vaquero, J.J.; Desco, M.; Muñoz-Barrutia, A.; Ripoll, J. Applications of light-sheet microscopy in microdevices. *Front. Neuroanat.* **2019**, *13*, 1. [CrossRef]
16. Paiè, P.; Bragheri, F.; Bassi, A.; Osellame, R. Selective plane illumination microscopy on a chip. *Lab A Chip* **2016**, *16*, 1556–1560. [CrossRef]
17. Regmi, R.; Mohan, K.; Mondal, P.P. Light sheet based imaging flow cytometry on a microfluidic platform. *Microsc. Res. Tech.* **2013**, *76*, 1101–1107. [CrossRef]
18. Lin, M.; Liu, Q.; Liu, C.; Qiao, X.; Shao, C.; Su, X. Label-free light-sheet microfluidic cytometry for the automatic identification of senescent cells. *Biomed. Opt. Express* **2018**, *9*, 1692–1703. [CrossRef]
19. Fan, Y.-J.; Hsieh, H.-Y.; Tsai, S.-F.; Wu, C.-H.; Lee, C.-M.; Liu, Y.-T.; Lu, C.-H.; Chang, S.-W.; Chen, B.-C. Microfluidic channel integrated with a lattice lightsheet microscopic system for continuous cell imaging. *Lab A Chip* **2021**, *21*, 344–354. [CrossRef]
20. Sala, F.; Castriotta, M.; Paiè, P.; Farina, A.; D'Annunzio, S.; Zippo, A.; Osellame, R.; Bragheri, F.; Bassi, A. High-throughput 3d imaging of single cells with light-sheet fluorescence microscopy on chip. *Biomed. Opt. Express.* **2020**, *11*, 4397–4407. [CrossRef]

21. Jiang, H.; Zhu, T.; Zhang, H.; Nie, J.; Guan, Z.; Ho, C.-M.; Liu, S.; Fei, P. Droplet-based light-sheet fluorescence microscopy for high-throughput sample preparation, 3-d imaging and quantitative analysis on a chip. *Lab A Chip* **2017**, *17*, 2193–2197. [CrossRef] [PubMed]
22. Behrouzi, M.; Rahimpouresfahani, F.; Youssef, K.; Rezai, P.; Tabatabaei, N. Optofluidic device for light-sheet fluorescence microscopy of c. elegans with a conventional wide-field microscope. In Proceedings of the Imaging, Manipulation, and Analysis of Biomolecules, Cells, and Tissues XX, San Francisco, CA, USA, 3 March 2022; Volume 11964, p. 1196402.
23. Memeo, R.; Paiè, P.; Sala, F.; Castriotta, M.; Guercio, C.; Vaccari, T.; Osellame, R.; Bassi, A.; Bragheri, F. Automatic imaging of drosophila embryos with light sheet fluorescence microscopy on chip. *J. Biophotonics* **2021**, *14*, e202000396. [CrossRef]
24. Vanwalleghem, G.; Schuster, K.; Taylor, M.A.; Favre-Bulle, I.A.; Scott, E.K. Brain-wide mapping of water flow perception in zebrafish. *J. Neurosci.* **2020**, *40*, 4130–4144. [CrossRef] [PubMed]
25. Hedde, P.N.; Malacrida, L.; Ahrar, S.; Siryaporn, A.; Gratton, E. sidespim–selective plane illumination based on a conventional inverted microscope. *Biomed. Opt. Express* **2017**, *8*, 3918–3937. [CrossRef] [PubMed]
26. Vogg, M.C.; Galliot, B.; Tsiairis, C.D. Model systems for regeneration: *Hydra*. *Dev.* **2019**, *146*, dev177212. [CrossRef] [PubMed]
27. Kier, W.M. The diversity of hydrostatic skeletons. *J. Exp. Biol.* **2012**, *215*, 1247–1257. [CrossRef]
28. Han, S.; Taralova, E.; Dupre, C.; Yuste, R. Comprehensive machine learning analysis of *Hydra* behavior reveals a stable basal behavioral repertoire. *Elife* **2018**, *7*, e32605. [CrossRef]
29. Carter, J.A.; Hyland, C.; Steele, R.E.; Collins, E.-M.S. Dynamics of mouth opening in *Hydra*. *Biophys. J.* **2016**, *110*, 1191–1201. [CrossRef]
30. Campbell, R.D. Structure of the mouth of *Hydra* spp. a breach in the epithelium that disappears when it closes. *Cell Tissue Res.* **1987**, *249*, 189–197. [CrossRef]
31. Iachetta, R.; Ambrosone, A.; Klimovich, A.; Wittlieb, J.; Onorato, G.; Candeo, A.; D'andrea, C.; Intartaglia, D.; Scotti, N.; Blasio, M.; et al. Real time dynamics of β-catenin expression during *Hydra* development, regeneration and wnt signalling activation. *Int. J. Dev. Biol.* **2018**, *62*, 311–318. [CrossRef]
32. Maroudas-Sacks, Y.; Garion, L.; Shani-Zerbib, L.; Livshits, A.; Braun, E.; Keren, K. Topological defects in the nematic order of actin fibres as organization centers of *Hydra* morphogenesis. *Nat. Phys.* **2021**, *17*, 251–259. [CrossRef]
33. Badhiwala, K.N.; Gonzales, D.L.; Vercosa, D.G.; Avants, B.W.; Robinson, J.T. Microfluidics for electrophysiology, imaging, and behavioral analysis of *Hydra*. *Lab A Chip* **2018**, *18*, 2523–2539. [CrossRef]
34. Badhiwala, K.N.; Primack, A.S.; Juliano, C.E.; Robinson, J.T. Multiple neuronal networks coordinate *Hydra* mechanosensory behavior. *Elife* **2021**, *10*, e64108. [CrossRef]
35. Edelstein, A.D.; Tsuchida, M.A.; Amodaj, N.; Pinkard, H.; Vale, R.D.; Stuurman, N. Advanced methods of microscope control using μmanager software. *J. Biol. Methods* **2014**, *1*, e10. [CrossRef]
36. Guan, Z.; Lee, J.; Jiang, H.; Dong, S.; Jen, N.; Hsiai, T.; Ho, C.-M.; Fei, P. Compact plane illumination plugin device to enable light sheet fluorescence imaging of multi-cellular organisms on an inverted wide-field microscope. *Biomed. Opt. Express* **2016**, *7*, 194–208. [CrossRef]
37. Steele, R.E.; Updegrove, M.D.; Kirolos, S.A.; Mowery, L.; Martinez, D.E.; Bryant, P.J. Reproductive bet-hedging and existence in vernal pools as components of *Hydra* life history. *Biol. Bull.* **2019**, *237*, 111–118. [CrossRef]
38. Trembley, A. *Mémoires Pour Servir à L'histoire D'un Genre de Ploypes D'eau Douce, à bras en Forme de Cornes*; Durand: Lower Beechmont, Australia, 1744; Volume 1.
39. Wilson, E.B. The heliotropism of *Hydra*. *Am. Nat.* **1891**, *25*, 413–433. [CrossRef]
40. Haug, G. Die lichtreaktionen der hydren. *Z. Für Vgl. Physiol.* **1933**, *19*, 246–303. [CrossRef]
41. Singer, R.H.; Rushforth, N.B.; Burnett, A.L. The photodynamic action of light on *Hydra*. *J. Exp. Zool.* **1963**, *154*, 169–173. [CrossRef]
42. Nath, T.; Mathis, A.; Chen, A.C.; Patel, A.; Bethge, M.; Mathis, M.W. Using deeplabcut for 3d markerless pose estimation across species and behaviors. *Nat. Protoc.* **2019**, *14*, 2152–2176. [CrossRef]
43. Gilpin, W.; Prakash, V.N.; Prakash, M. Flowtrace: Simple visualization of coherent structures in biological fluid flows. *J. Exp. Biol.* **2017**, *220*, 3411–3418. [CrossRef] [PubMed]
44. Gilpin, W.; Prakash, V.N.; Prakash, M. Dynamic vortex arrays created by starfish larvae. *Phys. Rev. Fluids* **2017**, *2*, 090501. [CrossRef]
45. Rodrigues, M.; Leclère, P.; Flammang, P.; Hess, M.W.; Salvenmoser, W.; Hobmayer, B.; Ladurner, P. The cellular basis of bioadhesion of the freshwater polyp *Hydra*. *BMC Zool.* **2016**, *1*, 3. [CrossRef]
46. Khetan, N.; Maheshwari, S.; Athale, C.A.A. Quantitative detachment mechanics of *Hydra* from substrates. *Proc. Indian Natn. Sci. Acad.* **2019**, *85*, 291–299. [CrossRef]
47. Shen, Y.; Siryaporn, A.; Lecuyer, S.; Gitai, Z.; Stone, H.A. Flow directs surface-attached bacteria to twitch upstream. *Biophys. J.* **2012**, *103*, 146–151. [CrossRef]
48. Kim, A.A.; Nekimken, A.L.; Fechner, S.; O'Brien, L.E.; Pruitt, B.L. Microfluidics for mechanobiology of model organisms. In *Methods in Cell Biology*; Elsevier: Amsterdam, The Netherlands, 2018; Volume 146, pp. 217–259.
49. Hwang, H.; Lu, H. Microfluidic tools for developmental studies of small model organisms–nematodes, fruit flies, and zebrafish. *Biotechnol. J.* **2013**, *8*, 192–205. [CrossRef]
50. Krishnamurthy, D.; Li, H.; Rey, F.B.; Cambournac, P.; Larson, A.G.; Li, E.; Prakash, M. Scale-free vertical tracking microscopy. *Nat. Methods* **2020**, *17*, 1040–1051. [CrossRef]

51. Hol, F.J.H.; Lambrechts, L.; Prakash, M. Biteoscope, an open platform to study mosquito biting behavior. *Elife* **2020**, *9*, e56829. [CrossRef]
52. Kumar, S.; Hol, F.J.H.; Pujhari, S.; Ellington, C.; Narayanan, H.V.; Li, H.; Rasgon, J.L.; Prakash, M. A microfluidic platform for highly parallel bite by bite profiling of mosquito-borne pathogen transmission. *Nat. Commun.* **2021**, *12*, 6018. [CrossRef]

biology

MDPI

Article

Analysis of Vibration Frequency and Direction for Facilitating Upper-Limb Muscle Activity

Cheng-Hua Ni [1,†], Yueh-Hsun Lu [2,3,†], Li-Wei Chou [4], Shu-Fen Kuo [5], Chia-Huei Lin [6], Shang-Lin Chiang [7], Liang-Hsuan Lu [4], Xin-Miao Wang [8], Jia-Lan Chang [9] and Chueh-Ho Lin [10,11,12,*]

1 Department of Nursing, Center for Nursing and Healthcare Research in Clinical Practice Application, Wan Fang Hospital, School of Nursing, College of Nursing, Taipei Medical University, Taipei 11031, Taiwan, R.O.C.
2 Department of Radiology, Shuang-Ho Hospital, Taipei Medical University, New Taipei City 235041, Taiwan, R.O.C.
3 Department of Radiology, School of Medicine, College of Medicine, Taipei Medical University, Taipei 11031, Taiwan, R.O.C.
4 Department of Physical Therapy and Assistive Technology, National Yang Ming Chiao Tung University, Taipei 112304, Taiwan, R.O.C.
5 School of Nursing, College of Nursing, Taipei Medical University, Taipei 11031, Taiwan, R.O.C.
6 Department of Nursing, Tri-Service General Hospital, School of Nursing, National Defense Medical Center, Taipei 114202, Taiwan, R.O.C.
7 Department of Physical Medicine and Rehabilitation, Tri-Service General Hospital, School of Medicine, National Defense Medical Center, Taipei 114202, Taiwan, R.O.C.
8 Faculty of Humanities, Zhejiang Dong Fang Polytechnic College, Wenzhou 325011, China
9 Department of Physical Medicine and Rehabilitation, Shuang Ho Hospital, Taipei Medical University, Taipei 235041, Taiwan, R.O.C.
10 Master Program in Long-Term Care, College of Nursing, Taipei Medical University, Taipei 11031, Taiwan, R.O.C.
11 International Ph.D. Program in Gerontology and Long-Term Care, College of Nursing, Taipei Medical University, Taipei 11031, Taiwan, R.O.C.
12 Center for Nursing and Healthcare Research in Clinical Practice Application, Wan Fang Hospital, Taipei Medical University, Taipei 116079, Taiwan, R.O.C.
* Correspondence: chueh.ho@tmu.edu.tw; Tel.: +886-2-2736-1661 (ext. 6325)
† These authors contributed equally to this study.

Simple Summary: Health promotion is important in older adults. However, aging-induced frailty results in poor muscle activity in the upper limbs, leading to activity impairments. Fortunately, recent studies report that vibration is a safe approach for improving muscular function; however, different frequencies and directions of vibrations can result in inconsistencies in muscle function improvement, and further investigation is needed. We developed a handheld vibrator to determine the effect of vibration frequency and direction on upper-limb muscle activation. Nineteen qualified participants were exposed to vertical and horizontal vibrations with 0, 15, 30, 45, and 60 Hz frequencies for 60 s each. Surface electromyography measured the activities of the flexor digitorum superficialis (FDS), flexor carpi radialis (FCR), extensor carpi ulnaris (ECU), extensor carpi radialis (ECR), biceps, triceps, and deltoid anterior muscles. Muscle activity was significantly induced under vibration conditions in both vertical and horizontal directions. The 45-Hz horizontal vibration induced maximum muscle activations for the FDS, ECR, ECU, biceps, and triceps. The 60-Hz vertical and 30-Hz horizontal vibrations facilitated maximum muscle activations for the FCR and deltoid anterior, respectively. We therefore suggest different protocols of vibration for specific weak muscles to improve muscle function in the upper limbs of older adults.

Abstract: We aimed to determine the effect of vibration frequency and direction on upper-limb muscle activation using a handheld vibrator. We recruited 19 healthy participants who were instructed to hold a handheld vibrator in their dominant hand and maintain the elbow at 90° flexion, while vertical and horizontal vibrations were applied with frequencies of 15, 30, 45, and 60 Hz for 60 s each. Surface electromyography (EMG) measured the activities of the flexor digitorum superficialis (FDS), flexor carpi radialis (FCR), extensor carpi ulnaris (ECU), extensor carpi radialis (ECR), biceps, triceps, and deltoid anterior muscles. EMG changes were evaluated as the difference in muscle activity between vibration and no-vibration (0 Hz) conditions. Muscle activity was induced under vibration conditions in both vertical and horizontal ($p < 0.05$) directions. At 45 Hz, FDS and FCR activities increased during horizontal vibrations, compared with those during vertical vibrations. ECU activity significantly increased under 15-Hz vertical vibrations compared with that during horizontal vibrations. Vibrations from the handheld vibrator significantly induced upper-limb muscle activity. The maximum muscle activations for FDS, ECR, ECU, biceps, and triceps were induced by 45-Hz horizontal vibration. The 60-Hz vertical and 30-Hz horizontal vibrations facilitated maximum muscle activations for the FCR and deltoid anterior, respectively.

Keywords: vibration; direction; frequency; muscle; electromyography; upper limb

1. Introduction

Muscle activation and related upper-limb functions play important roles in performing activities of daily living such as feeding, bathing, and dressing. Aging or neurologic disorders may result in upper-limb motor control and functional impairment [1]. Many studies have indicated that older adults with degenerative conditions exhibit significant motor function deficits, which lead to frailty and long-term disabilities [2–4]. However, studies have also revealed that the frailty could result from the physiologic age-dependent changes in vicious loops [5,6] or lifestyle (sedentary behaviors) [7], leading to poor muscle strength and muscle atrophy (sarcopenia) [5,6]. This could result in muscle weakness, low physical activity, and impaired upper-limb function [5–13] and could increase dependency levels in daily living for older persons. Moreover, older adults are reportedly predisposed to developing muscle atrophy and weakness. Greenlund et al. reported that mortality due to stroke—one of the most common neurologic diseases in older adults—has decreased [14]; however, stroke-induced motor function deficits result in paretic upper limbs [15,16], which alter the daily functioning and quality of life and cause long-term disabilities in patients with stroke [17,18]. Only 15% of patients in the acute and subacute stages of stroke who undergo traditional rehabilitation interventions recover their paretic upper-limb motor function [19]. An appropriate rehabilitation intervention is required to improve upper-limb neuromuscular function and motor recovery in older adults and people after stroke.

The vibration approach is reportedly effective, easy, satisfactory, and safe for improving muscle strength, power, and function [20–23]. The mechanism for muscle function improvement is based on facilitating the activation of the efferent Ia muscle fibers during vibration approaches, leading to α-motor neuron excitation, which produces more muscle force output [24,25]. The vibration approach included asking the participants to sit on a chair or kneel on the ground and to place their hands or elbows on the vibratory platform [22,26–32] or to directly sit on the vibratory platform with two hands in the sideways position on the vibratory platform [33]. Studies have shown that different vibration approach protocols (frequencies between 5 and 60 Hz) and vibration directions (horizontal or vertical) lead to improvements in muscle strength [25,34–36]. However, contradictory findings were also reported in recent studies [22,37]. Vibration transmission is a complex process that is influenced by biomechanics [27,38]; beside the vibration frequencies, the different postures during vibration approaches may also affect upper-limb muscle activation [22,27]. This indicates that the force translation to the upper limbs may change because different postures may result in different vibration directions from the vibrator. Based on previous findings regarding the mechanism and inconsistent findings of vibration approaches, we speculated that vibration frequency and directions constitute key factors that facilitate muscle activation during vibration approaches. In addition, the outcome of vibration application on the upper-limb musculature remains inconclusive because of differences in the vibration stimuli applied and limited appropriate equipment; only a few studies have applied vibration-induced muscle activations in the upper limbs [27,28,39–41].

Furthermore, many studies have suggested that the efficacy of appropriate equipment with optimal vibration protocols to induce muscle activation should be established before applying these protocols clinically [23–25,42–45]. Furthermore, recent studies have suggested that vibration-induced changes in neuromuscular activation during different vibration stimuli can be evaluated directly using the root mean squares of surface electromyographic (EMG) signals (EMGrms), which reveal the effect of vibration stimuli on muscle strength enhancement [27]. Therefore, this study aimed to develop a handheld vibrator to determine the effect of vibration frequency and direction on upper-limb muscle activation.

2. Material and Methods

2.1. Participants

The sample size requirement was calculated using G*power (version 3.1.9.2, Heinrich-Heine-Universität, Düsseldorf, Germany) using an effect size of 0.25, an alpha of 0.05, and a power of 0.80. A total sample size of 34 participants was required. However, due to the coronavirus disease 2019 (COVID-19) pandemic, only 19 adults were recruited. The inclusion criteria required that participants be able to follow the researcher's instructions and study procedures, be healthy with no cognitive disorders that could affect the vibration approach performance, be able to steadily hold a handheld vibrator using the dominant hand without pain or discomfort, and have good cognitive function. We excluded patients with acute or chronic neurologic or orthopedic impairments and those who experienced discomfort or had undergone surgery in the upper limbs within 6 months prior to the study onset. The dominant hand for each participant was considered the hand that was used to sign the informed consent form.

This study was conducted in communities and approved by the institutional review board of Taipei Medical University (approval number: N202007048). All participants signed an informed consent form before participating in this study. This study was carried out in accordance with the Declaration of Helsinki.

2.2. Research Device and Data Processing

A handheld vibrator (measuring 37 × 8.3 × 10.5 cm; weight, 1 kg) was designed and customized by ACCU BALANCES CORP. (New Taipei City, Taiwan). The handheld vibrator motor (ZYT3424D110; 1.833 V = 1 Hz) produced vibrations with a 5-mm amplitude and 0–60-Hz frequency range, and its validity was confirmed by analyzing the vibration frequencies generated by the vibrator between 0–60-Hz with APDM-collected vibration frequency data. The vibration frequency could be controlled using a vibration controller (rotary potentiometer) located on the right side of the vibration frequency control box; a frequency mechanical pointer located on the left side of the box showed the voltage applied to the vibration motor, which is correlated with the frequency measurement via APDM (Figure 1). To determine the effects of vibration on the upper limbs, a red line target indicated the position for holding the vibrator during vertical (up and down vibration in the sagittal plane) and horizontal (medial and lateral vibration in the horizontal plane) vibration approaches (Figure 2). The handheld vibrator was applied to generate vibration force to the entire upper limb rather than a specific muscle during the vibration tests.

The research device includes a handheld vibrator and vibration frequency control box. The researcher controls the vibration frequency, which is displayed on the mechanical pointer, using the vibration controller.

Participants used their dominant hand to hold the vibrator at the red line. Participants held the vibrator in two different ways (Figure 2a,b) so that the direction of vibration was either up–down or lateral–medial.

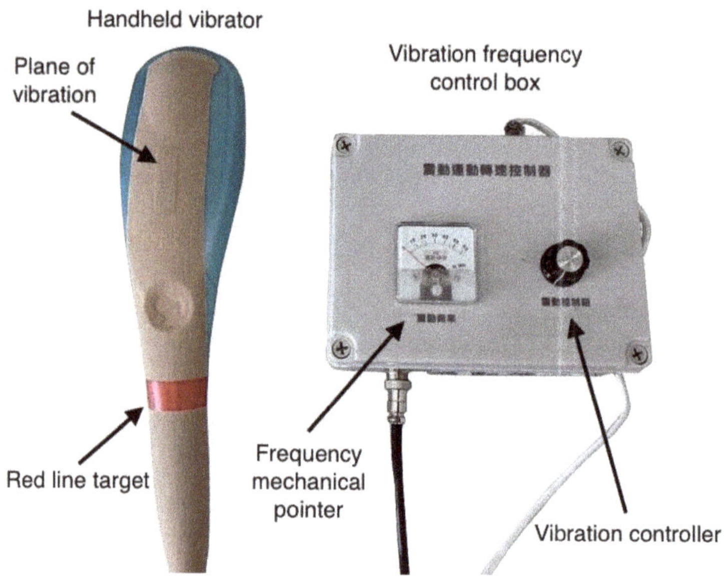

Figure 1. Handheld vibrator and related component parts.

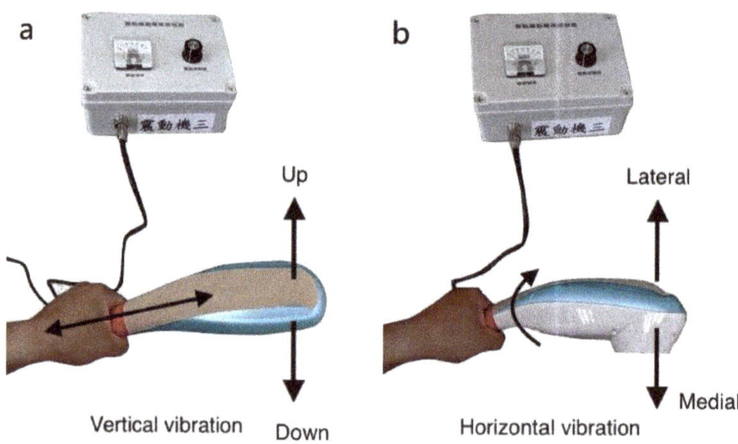

Figure 2. Illustrations for vertical (**a**) and horizontal (**b**) vibration conditions.

2.3. Experimental Procedures and Positioning of Participants

All participants were instructed to sit back on a high-fixed, no-arm support chair with their feet positioned flat on the floor. The dominant shoulder was positioned sideways, slightly apart from the trunk, and the elbow was fixed at 90° flexion as the standard vibration position (Figure 3). Thereafter, the researcher instructed the participants to hold the red line on the vibrator firmly with the dominant hand. Participants randomly performed all vibration tests comprising five vibration frequencies (0, 15, 30, 45, and 60 Hz) in the vertical and horizontal vibration directions. Moreover, participants were instructed to perform all vibration approaches while maintaining the elbow in the standard vibration position.

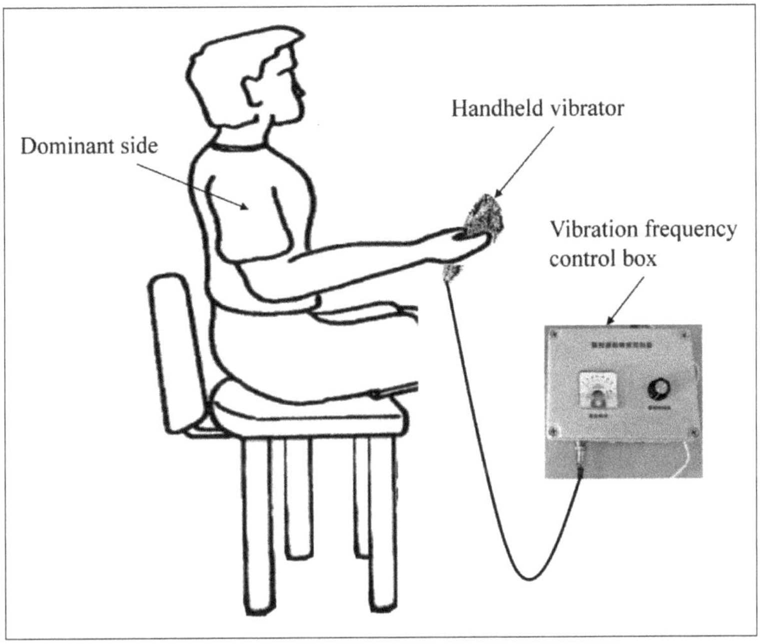

Figure 3. Experimental setting for the vibration.

Based on our review of previous clinical studies on improving muscle strength, we used vibration protocols with vibration exposure between 30 and 60 s and with resting intervals between 15 and 60 s [46–49]. Studies have reported that fatigue [50], muscle adaption, and injury risk [24] can increase when vibrations last more than 1 min. Previous research also reported that the vibration approach has 30 s of maintained effects on facilitating muscle excitation [51]. Therefore, the vibration approach was performed for 60 s with a 1-min resting interval between sessions to avoid muscle fatigue.

2.4. EMG Analysis

Seven muscle groups, including the flexor digitorum superficialis (FDS), flexor carpi radialis (FCR), extensor carpi radialis (ECR), extensor carpi ulnaris (ECU), biceps, triceps, and deltoid anterior, were selected to represent the performance of the upper limbs [22,52]. EMG signals were measured using the BTS FREEEMG 1000 with EMG-BTS EMG-Analyzer® (BTS Bioengineering, Milan, Italy) during vibration exposure, at a sampling frequency of 1000 Hz. Based on the Surface Electromyography for the Non-invasive Assessment of Muscles guidelines, pairs of bipolar Ag-AgCl electrodes (H124SG Covidien, Minneapolis, MN, USA) were placed over the belly of each muscle, with an inter-electrode distance of 2 cm [53]. Before placing the electrodes, each participant's skin was thoroughly cleaned with alcohol swabs [27]. Electromyography (EMG) signals were amplified with a gain of 1000. EMG post-processing was performed using the EMG-Analyzer. We used bandpass EMG signals (20–400 Hz); a notch filter was used to remove the noise from the power line (60 Hz). EMGrms, with a 100-ms window, was used to process EMG data for each muscle during five vibration approaches sessions [27,30,39,54]. Further, the EMGrms values were normalized with respect to the percentage of maximum voluntary contraction of each corresponding muscle for each participant [39].

2.5. Statistical Analysis

The Shapiro–Wilk test was performed to test the normality of the sample data. The Mann–Whitney U test was performed to confirm whether the vibration approaches had a significant impact on individual muscle activation in the vertical and/or horizontal vibration directions. Further, the Friedman test was performed to compare the increase in muscle activation between vibration (15, 30, 45, and 60 Hz) and no-vibration (0 Hz) conditions. Lastly, the Wilcoxon signed-rank test was performed to compare vibration direction-related changes in muscle activation for each muscle with different vibration frequencies in the vertical and horizontal vibration directions. The alpha level was set at 0.05. Statistical Package for the Social Sciences software (version 17.0, SPSS Inc., Chicago, IL, USA) was used for statistical analysis.

3. Results

3.1. Participants

Nineteen adults (age 38.2 ± 14.0 years; 14 female and 5 male) were recruited in this study (Table 1).

Table 1. Demographic characteristics.

	Participants (N = 19)
Age (years)	38.2 ± 14.0
Body weight (kg)	64.3 ± 13.4
Body height (cm)	163.8 ± 9.9
Sex (F/M)	14/5
Dominant side (R/L)	17/2

Data are expressed as mean ± standard deviation or number. Abbreviations: F, female; M, male; R, right; L, left.

3.2. EMG Analysis

All variables were assumed to be non-normally distributed (all $p < 0.001$, respectively), including FDS, FCR, ECR, ECU, biceps, triceps, and deltoid anterior.

On comparing the vibration direction-related changes in muscle activation for upper-arm and shoulder muscles, the results showed no statistically significant differences in muscle activation for the biceps, triceps, and deltoid anterior muscles at 0, 15, 30, 45, and 60 Hz vibration frequencies in the horizontal and vertical vibration directions (Table 2; Figure 4).

Table 2. Vibration-induced changes in EMGrms (%) compared with those in the no-vibration condition.

	Vertical Vibration [b]			Horizontal Vibration [b]			
	Mean (SD)	Wilcoxon Signed-Rank Test [c]	Friedman Test [b]	Mean (SD)	Wilcoxon Signed-Rank Test [c]	Friedman Test [b]	Mann–Whitney U Test [a]
		Z-Value (p Value)	X^2-Value (p Value)		Z-Value (p Value)	X^2-Value (p Value)	Z-Value (p Value)
Flexor digitorum superficialis			35.07 (0.000 [**b])			46.82 (0.000 [**b])	
0 Hz	10.92 (6.58)	-		9.44 (5.69)	-		158.50 (0.521 [a])
15 Hz	29.26 (17.91)	−3.783 (0.000 [**c])		26.95 (17.16)	−3.662 (0.000 [**c])		166.00 (0.672 [a])
30 Hz	25.61 (14.00)	−3.743 (0.000 [**c])		32.28 (18.62)	−3.783 (0.000 [**c])		217.00 (0.287 [a])
45 Hz	25.01 (13.73)	−3.421 (0.000 [**c])		38.01 (18.54)	−3.823 (0.000 [**c])		261.00 (0.019 [*a])
60 Hz	27.76 (17.79)	−3.421 (0.000 [**c])		35.17 (18.23)	−3.783 (0.000 [**c])		223.00 (0.215 [a])
Flexor carpi radialis			43.41 (0.000 [**b])			34.31 (0.000 [**b])	
0 Hz	11.32 (4.83)	-		13.07 (8.70)	-		188.00 (0.827 [a])
15 Hz	33.21 (17.15)	−3.823 (0.000 [**c])		35.92 (19.38)	−3.461 (0.000 [**c])		190.00 (0.782 [a])
30 Hz	31.69 (17.81)	−3.783 (0.000 [**c])		44.23 (26.75)	−3.501 (0.000 [**c])		234.00 (0.118 [a])
45 Hz	31.52 (17.20)	−3.823 (0.000 [**c])		49.40 (29.40)	−3.622 (0.000 [**c])		253.00 (0.034 [*a])
60 Hz	37.66 (20.64)	−3.823 (0.000 [**c])		49.20 (32.62)	−3.541 (0.000 [**c])		222.00 (0.226 [a])
Extensor carpi radialis			39.70 (0.000 [**b])			59.95 (0.000 [**b])	
0 Hz	14.76 (7.40)	-		15.74 (12.68)	-		175.00 (0.872 [a])
15 Hz	53.20 (34.15)	−3.823 (0.000 [**c])		38.83 (21.39)	−3.823 (0.000 [**c])		136.00 (0.194 [a])
30 Hz	48.80 (28.18)	−3.783 (0.000 [**c])		48.03 (23.24)	−3.823 (0.000 [**c])		182.00 (0.965 [a])
45 Hz	49.5 6 (24.13)	−3.823 (0.000 [**c])		60.47 (28.83)	−3.823 (0.000 [**c])		222.00 (0.226 [a])
60 Hz	55.50 (29.73)	−3.823 (0.000 [**c])		59.85 (30.65)	−3.823 (0.000 [**c])		194.00 (0.693 [a])
Extensor carpi ulnaris			38.82 (0.000 [**b])			49.72 (0.000 [**b])	
0 Hz	11.83 (5.64)	-		9.83 (5.05)	-		145.50 (0.307 [a])
15 Hz	38.97 (22.46)	−3.823 (0.000 [**c])		22.95 (11.22)	−3.823 (0.000 [**c])		101.00 (0.020 [*a])

Table 2. *Cont.*

	Vertical Vibration [b]			Horizontal Vibration [b]			
	Mean (SD)	Wilcoxon Signed-Rank Test [c]	Friedman Test [b]	Mean (SD)	Wilcoxon Signed-Rank Test [c]	Friedman Test [b]	Mann–Whitney U Test [a]
		Z-Value (p Value)	X^2-Value (p Value)		Z-Value (p Value)	X^2-Value (p Value)	Z-Value (p Value)
30 Hz	34.21 (18.26)	−3.823 (0.000 **c)		26.64 (14.94)	−3.823 (0.000**c)		133.00 (0.166 a)
45 Hz	34.91 (23.07)	−3.823 (0.000 **c)		34.25 (22.56)	−3.823(0.000**c)		183.00 (0.942 a)
60 Hz	35.90 (29.90)	−3.702 (0.000 **c)		31.39 (20.81)	−3.823 (0.000**c)		173.00 (0.827 a)
Biceps			54.61 (0.000 **b)			53.72 (0.000 **b)	
0 Hz	7.22 (3.92)	-		7.37 (3.57)	-		191.50 (0.748 a)
15 Hz	21.89 (10.69)	−3.823 (0.000 **c)		21.56 (7.64)	−3.823(0.000 **c)		185.00 (0.895 a)
30 Hz	27.12 (13.23)	−3.823 (0.000 **c)		29.24 (12.58)	−3.823(0.000 **c)		202.00 (0.530 a)
45 Hz	32.21 (15.63)	−3.823 (0.000 **c)		41.18 (29.81)	−3.823(0.000 **c)		202.00 (0.530 a)
60 Hz	33.38 (16.27)	−3.823 (0.000 **c)		34.64 (17.89)	−3.823(0.000 **c)		189.00 (0.804 a)
Triceps			35.74 (0.000 **b)			48.46 (0.000 **b)	
0 Hz	3.19 (2.02)	-		3.74 (3.00)	-		191.50 (0.748 a)
15 Hz	9.94 (6.11)	−3.823 (0.000 **c)		8.89 (7.30)	−3.823 (0.000 **c)		157.00 (0.493 a)
30 Hz	9.35 (5.39)	−3.743 (0.000 **c)		9.27 (6.34)	−3.823 (0.000 **c)		168.00 (0.715 a)
45 Hz	11.17 (9.14)	−3.662 (0.000 **c)		11.64 (7.88)	−3.823 (0.000 **c)		192.00 (0.737 a)
60 Hz	13.27 (11.54)	−3.823 (0.000 **c)		12.62 (10.05)	−3.823 (0.000 **c)		176.00 (0.895 a)
Deltoid anterior			39.11 (0.000 **b)			43.91 (0.000 **b)	
0 Hz	8.01 (8.14)	-		12.72 (11.66)	-		246.50 (0.054 a)
15 Hz	31.36 (32.06)	−3.823 (0.000 **c)		43.04 (34.16)	−3.823 (0.000 **c)		241.00 (0.077 a)
30 Hz	30.46 (23.82)	−3.823 (0.000 **c)		43.79 (27.79)	−3.823 (0.000 **c)		240.00 (0.082 a)
45 Hz	28.91 (19.48)	−3.823 (0.000 **c)		43.55 (28.83)	−3.823 (0.000 **c)		237.00 (0.099 a)
60 Hz	32.99 (25.55)	−3.823 (0.000 **c)		40.16 (32.46)	−3.702 (0.000 **c)		205.00 (0.474 a)

* Significant difference $p < 0.05$. ** Significant difference $p < 0.01$. [a] Mann–Whitney U test was performed; [b] Friedman test was performed; [c] Wilcoxon signed-rank test was performed; significantly greater than that in the no-vibration (0 Hz) condition. Vibrations were applied at frequencies of 15, 30, 45, and 60 Hz in the vertical and horizontal directions. EMGrms, root mean square of surface electromyographic signals; SD, standard deviation.

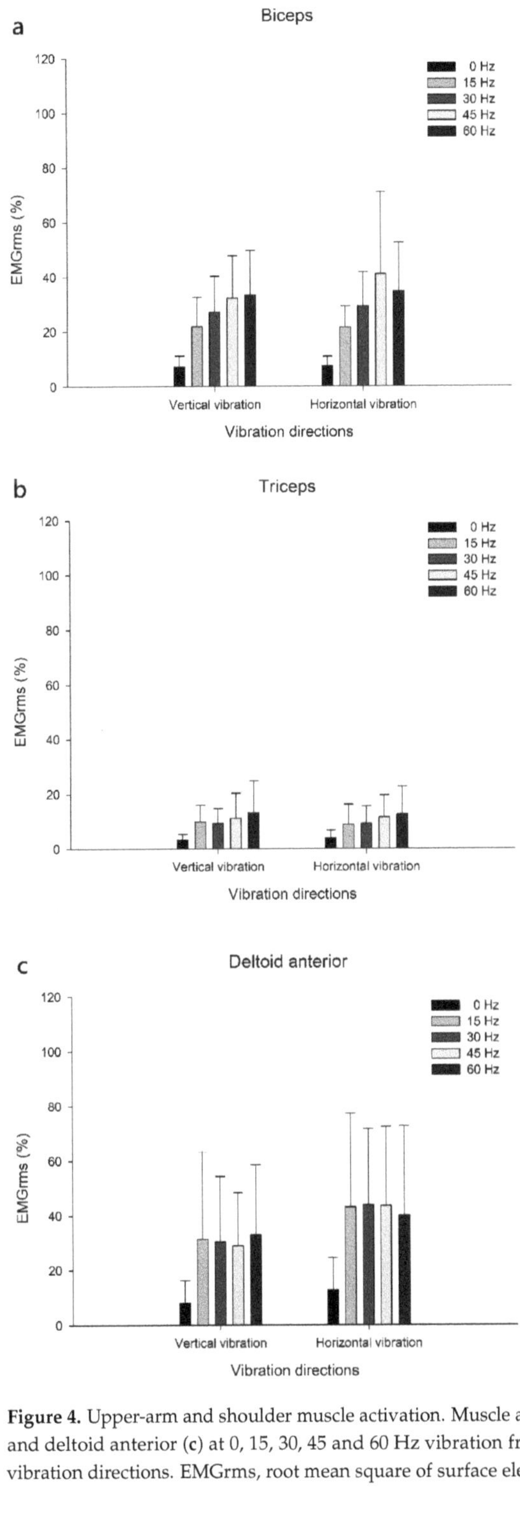

Figure 4. Upper-arm and shoulder muscle activation. Muscle activations in the biceps (**a**), triceps (**b**), and deltoid anterior (**c**) at 0, 15, 30, 45 and 60 Hz vibration frequencies for vertical and horizontal vibration directions. EMGrms, root mean square of surface electromyographic signals.

However, the FDS and FCR muscles were more activated during the 45-Hz vibration approach in the horizontal vibration direction than in the vertical direction ($p < 0.05$). In addition, the ECU muscles were more activated during the 15-Hz vibration approach in the vertical vibration direction than in the horizontal direction (Table 2; Figure 5).

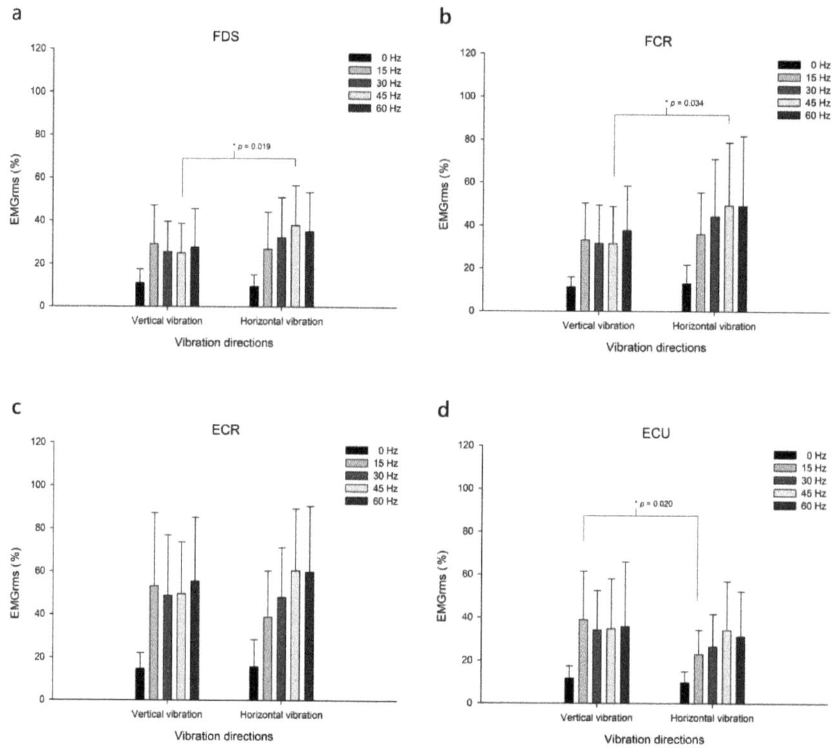

Figure 5. Forearm muscle activation. Muscle activations in the FDS (**a**), FCR (**b**), ECR (**c**), and ECU (**d**) muscles of the forearm at 0, 15, 30, 45, and 60 Hz vibration frequencies in the vertical and horizontal vibration directions. EMGrms, root mean square of surface electromyographic signals. FDS, flexor digitorum superficialis; FCR, flexor carpi radialis; ECR, extensor carpi radialis; ECU, extensor carpi ulnaris.

The vibration approaches significantly impacted all individual muscle activations in the vertical ($p < 0.05$) and horizontal ($p < 0.05$) vibration directions (Table 2).

All muscle groups had significantly facilitated muscle activation in the vibration condition compared with that in the no-vibration condition, in both vibration directions (Table 2). Furthermore, compared with no-vibration (0 Hz), for both vertical and horizontal vibration directions, the maximum muscle activation for FCR was facilitated by 60 Hz vertical vibration ($p < 0.001$); the maximum muscle activations for FDS, ECR, ECU, biceps, and triceps were induced during 45 Hz horizontal vibration ($p < 0.001$), respectively. For the deltoid anterior, the maximum muscle activation frequency was induced during 30 Hz horizontal vibration ($p < 0.001$) (Table 2).

4. Discussion

Many studies have used whole-body vibration devices to investigate the impact of vibration approaches on upper-limb muscle activation [22,26–29,31,32]. In contrast, this study developed a frequency-controlled handheld vibrator with a focus on upper-limb muscle activation. We found that vibration approaches with specific frequencies (15, 30, 45, and 60 Hz) had a positive effect on muscle activation in all seven muscle groups i.e., FDS, FCR, ECR, ECU, biceps, triceps, and deltoid anterior in both the vertical and horizontal directions, unlike in the absence of vibration (0 Hz). Furthermore, horizontal and vertical vibrations had a significant facilitatory effect on the activation of upper-limb flexors (FDS and FCR muscles) and extensors (ECU muscles), respectively. To the best of our knowledge, this is the first study to report that different vibration directions induce the activation of different upper-limb muscles.

4.1. Vibrator Types and Applications in Upper-Limb Muscle Activity Induction

Most previous studies used vibratory platforms as the vibrator, and participants were instructed to take specific positions for upper-limb muscle activation. For instance, participants were instructed to sit on a chair or kneel on the ground and then support their upper limbs over the vibratory platform with their hands or elbows shoulder-width apart [22,26–32]. In other studies, participants were instructed to hold the handrail on the top of the vibratory platform [27] or sit on the vibratory platform with two hands in the sideways position [33]. The approaches used in previous studies may or may not have induced upper-limb muscle activation; however, they could cause vertigo or discomfort as the inappropriate and excess vibration force from whole-body vibration could be transmitted to the head [21]. Approximately 2.4–3.6% of individuals who undergo whole-body vibration approaches develop vertigo [21], muscle soreness [21,55] and discomfort [21,24,56]. Few studies have applied flexi-bar approaches for upper-limb muscle strength enhancement; nonetheless, the vibration frequency and amplitude derived from flexi-bar approaches cannot be used for vibration approaches because flexi-bar vibrations are generated, and their frequencies change based on the participant's force and skill [57]. In this study, participants were instructed to hold the handheld vibrator in their dominant hand, with their trunk supported on a chair and to maintain their upper limbs in a specific vibration position. Unlike previous studies, our study found that the use of frequency-controlled vibrations from the handheld vibrator could significantly facilitate upper-limb muscle activation without the occurrence of adverse effects such as vertigo and discomfort.

4.2. Influence of Vibration Frequency on Upper-Limb Muscle Activation

We found that vibration approaches significantly facilitated upper-limb muscle activation in the vertical and horizontal vibration directions. Unlike no-vibration conditions, vibration conditions enhanced upper-limb muscle activation by 31.4–52.2% and 28.5–54.9% in the vertical and horizontal vibration directions, respectively. Specifically, the maximum muscle activation for FCR was facilitated at 60 Hz in the vertical vibration direction; the maximum muscle activations for FDS, ECR, ECU, biceps, and triceps were induced at 45 Hz in the horizontal vibration direction, respectively. For the deltoid anterior, the maximum muscle activation frequency was 30 Hz in the horizontal direction. Hence, vibration frequencies between 30 and 60 Hz may be used to facilitate maximum muscle activations, and more than half of the muscle groups were activated by high-frequency vibrations (45 and 60 Hz). Recent studies also reported similar findings and showed that a high-frequency vibration has a greater facilitatory effect on muscle activation than a low-frequency vibration in healthy participants and patients with stroke [39,43]. However, the increased muscle activation does not linearly correlate with increasing vibration frequencies [34]. For example, the activation of all muscles at 60 Hz increased the EMGrms by only 0.9–9% rather than doubling it in both vertical and horizontal vibration directions. This phenomenon may be the result of a damping reaction on muscle activation when performing vibration [34]. This corroborates the findings of a recent study, which reported that doubling the acceleration of the applied vibration only facilitates a 3–5% increase in EMGrms [58]. Our findings show that high-frequency vibrations facilitate upper-limb muscle activation, but they may lead to discomfort. For example, whole-body vibration can cause bone frailty, back pain [38,59], resonance injury, and dizziness [21,56]; a vibration frequency of

>40 Hz may significantly affect posture control, which results in muscle fatigue [60] and an unstable posture sway [61]. In this study, no participants reported any discomfort or dizziness during or after vibration testing. This may be because the participants were asked to perform all vibration approaches while maintaining the upper limbs in the standard vibration position. Most of the vibration force was thus absorbed by the upper limbs, with little vibration force translating to the body and head. If individuals are unable to hold the vibrator and resist the vibration force during high-frequency vibration approaches, injuries may ensue [62]. Therefore, based on the findings of this study along with those of previous studies, we suggest that a 30-Hz vibration frequency in combination with active muscle participation may be the optimal and safe vibration approach for upper-limb muscle activation. Further studies should be conducted to confirm the benefits of this frequency toward improving muscle strength and function for people with disabilities.

4.3. Vibration Direction Affects Upper-Limb Muscle Activation

The FDS and FCR activities at a 45-Hz vibration frequency during horizontal vibrations increased by 9.7% and 6.2%, respectively, compared to that during vertical vibrations. In addition, ECU activation at a 15-Hz vibration frequency was significantly higher in the vertical than in the horizontal vibration direction, by 11.7%. Horizontal vibration has a greater facilitatory effect on FDS muscle activation probably because more muscle activation is required to maintain the handheld vibrator in a fixed position. Additionally, we found that many participants unconsciously flexed their wrists to maintain stability during vibration, which could result in a greater facilitatory effect on FCR activation. Vertical vibration tends to produce a greater facilitatory effect on extensor muscle activation, probably because the device vibrates in the direction of gravity; hence, extensors resist both gravity and the vibration force from the vibrator to maintain the vibration approach posture. Different postures may also result in different vibration directions for the upper limbs during vibration approaches; nonetheless, few studies have investigated the impact of different postures on upper-limb muscle activation. Two recent studies recruited healthy participants to perform vibration approaches in different postures, such as standing on the vibratory platform with arms along the body (squat posture), holding a handrail on top of the vibratory platform in the half-squatting position, and kneeling on the ground with both hands supported over the platform (push-up modified posture) [22,27]. These studies found that muscle strength increases in the push-up modified posture than in the half-squatting posture [27], and the EMGrms for the FDS muscle during squatting was significantly higher than that in the non-vibration condition by 23.3% [22]. However, these findings are difficult to compare with those of the present study because the vibration approaches were different, warranting further investigation.

4.4. Study Limitations and Future Perspectives

Our study has some limitations. We developed a frequency-controlled handheld vibrator and indicated the facilitatory effect of vibration approaches on upper-limb muscle activation; however, vibration force transmission and the activation of different upper-limb muscles are complex processes that are influenced by biomechanics [27,38] and vibration protocol parameters (frequency, amplitude, displacement, vibration time, types of vibration, and postures) [42]. Our study only evaluated the effects of vibration frequency and direction on upper-limb muscle activation; hence, the impacts of the other vibration protocol parameters require further investigation. In addition, we suggest that future studies use wearable accelerometers to directly measure the vibration-induced accelerations in target muscle groups of the upper limbs, with analysis of their relationship to muscle activation. For participants, the ratio of male to female participants was not equal in this study. A recent study reported that the body mass of males is higher than that of females and found that vibration-induced accelerations of the body in three dimensions are lower in males [63]. This finding suggests that sex-related differences in body properties and structure may impact the effects of vibration on muscle activation and should thus be considered when applying the vibration approach clinically. Additionally, participants were asked to maintain a firm grip of the vibration device during the experiments. We believe the grip force was approximately constant throughout the experiment, otherwise the participants would have lost grip of the device. However, we did not measure the grip force or pressure underneath, which may have affected the amount of vibration delivered. Future studies could investigate the effects of grip force levels on muscle strength gain due to vibration training. We found many participants unconsciously flexed their wrists to maintain the vibration position of the upper limbs during vibration, which could have been caused by the posture response-induced muscle activation. Future studies should design appropriate vibrators related to the upper limbs, rather than using whole-body vibrators. Furthermore, APDM was applied to validate the vibration frequencies generated by the vibrator between 0 and 60 Hz in this study. However, the maximum sampling frequency for APDM is 128 Hz. To measure vibration frequencies of 60 Hz, a sampling rate that is at least 10 times higher (i.e., $\geq$600 Hz) is ideal. Theoretically, using a sampling rate that is twice the frequency seems sufficient, but that is the bare minimum. Future studies should use a rate that is at least 10 times larger to collect more accurate measurement of vibration frequency. Meanwhile, due to the COVID-19 pandemic, participant recruitment was limited. Future studies should include a large sample size, along with optimal vibration frequencies and directions during training programs. This would facilitate muscle activation and improve muscle function and functional recovery in frail, older adults, and patients with stroke.

5. Conclusions

In this study, we developed a handheld vibrator and validated its positive effects on upper-limb muscle activation. The 45-Hz horizontal vibration approach can induce maximum activations for the FDS, ECR, ECU, biceps, and triceps muscles. Moreover, 60-Hz vertical and 30-Hz horizontal vibrations can facilitate the maximum muscle activations for the FCR and deltoid anterior, respectively. Further studies should be conducted to confirm the benefits of this frequency toward improving muscle strength and function for people with disabilities.

Author Contributions: Conceptualization, C.-H.N., Y.-H.L., C.-H.L. (Chia-Huei Lin) and C.-H.L. (Chueh-Ho Lin); Data curation, S.-F.K., L.-H.L. and X.-M.W.; Formal analysis, S.-F.K., C.-H.L. (Chia-Huei Lin) and J.-L.C.; Funding acquisition, C.-H.N. and C.-H.L. (Chueh-Ho Lin); Investigation, C.-H.N., Y.-H.L., L.-W.C., C.-H.L. (Chia-Huei Lin), S.-L.C., L.-H.L., X.-M.W., J.-L.C. and C.-H.L. (Chueh-Ho Lin); Methodology, C.-H.N., Y.-H.L., L.-W.C., S.-F.K., C.-H.L. (Chia-Huei Lin), S.-L.C., L.-H.L., X.-M.W., J.-L.C. and C.-H.L. (Chueh-Ho Lin); Project administration, C.-H.N., Y.-H.L. and C.-H.L. (Chueh-Ho Lin); Resources, C.-H.N., Y.-H.L., L.-W.C., C.-H.L., S.-L.C., L.-H.L., X.-M.W., J.-L.C. and C.-H.L. (Chueh-Ho Lin); Software, L.-W.C. and S.-F.K.; Supervision, C.-H.N., Y.-H.L., L.-W.C., S.-L.C. and C.-H.L. (Chueh-Ho Lin); Validation, C.-H.N., Y.-H.L., L.-W.C., S.-F.K., S.-L.C., L.-H.L., X.-M.W., J.-L.C. and C.-H.L. (Chueh-Ho Lin); Visualization, C.-H.N. and L.-W.C.; Writing—original draft, C.-H.N., Y.-H.L., L.-W.C., S.-F.K., C.-H.L. (Chia-Huei Lin), S.-L.C., L.-H.L., X.-M.W. and C.-H.L. (Chueh-Ho Lin); Writing—review & editing, C.-H.L. (Chueh-Ho Lin). All authors have read and agreed to the published version of the manuscript.

Funding: This research was funded by the Ministry of Science and Technology (MOST) of Taiwan (MOST 111-2221-E-038-015, MOST 111-2622-E-038-002, MOST 110-2221-E-038-016, MOST 109-2622-B-038-007) and Taipei Medical University—Wan Fang Hospital (108TMU-WFH-24).

Institutional Review Board Statement: The study was conducted in accordance with the Declaration of Helsinki and approved by the Institutional Review Board of Taipei Medical University (Approval Number: N202007048, 5 August 2022).

Informed Consent Statement: Written informed consent has been obtained from the patient(s) in this study.

Data Availability Statement: The datasets are available from the corresponding author upon reasonable request.

Acknowledgments: The authors would like to thank all study participants, research assistant Chia-Hui Kao who conducted this study, and the Editage team for English language editing. We acknowledge the (statistical/computational/technical) support of the Health Data Analytics and Statistics Center, Office of Data Science, Taipei Medical University, Taiwan.

Conflicts of Interest: The authors declare no competing interest.

References

1. Shumway-Cook, A.; Woollacott, M.H. *Motor Control: Theory and Practical Applications*; Lippincqtt Williams & Wilkins: Philadelphia, PA, USA, 1995.
2. Roos, M.R.; Rice, C.L.; Vandervoort, A.A. Age-related changes in motor unit function. *Muscle Nerve* **1997**, *20*, 679–690. [CrossRef]
3. Desrosiers, J.; Hebert, R.; Bravo, G.; Rochette, A. Age-related changes in upper extremity performance of elderly people: A longitudinal study. *Exp. Gerontol.* **1999**, *34*, 393–405. [CrossRef] [PubMed]
4. Woo, J.; Yu, R.; Wong, M.; Yeung, F.; Wong, M.; Lum, C. Frailty Screening in the Community Using the FRAIL Scale. *J. Am. Med. Dir. Assoc.* **2015**, *16*, 412–419. [CrossRef] [PubMed]
5. Mühlberg, W.; Sieber, C. Sarcopenia and frailty in geriatric patients: Implications for training and prevention. *Z. Gerontol. Geriatr.* **2004**, *37*, 2–8. [CrossRef]
6. Seene, T.; Kaasik, P. Muscle weakness in the elderly: Role of sarcopenia, dynapenia, and possibilities for rehabilitation. *Eur. Rev. Aging Phys. Act.* **2012**, *9*, 109–117. [CrossRef]
7. Teixeira-Santos, L.; Bobrowicz-Campos, E.; Parola, V.; Coelho, A.; Gil, I.; Almeida, M.d.L.; Apóstolo, J.L. What Is the Relationship between Lifestyle and Frailty Status? Data from the Portuguese Multicentre Descriptive Study. *Nurs. Rep.* **2022**, *12*, 39–49. [CrossRef]
8. Shechtman, O.; Mann, W.C.; Justiss, M.D.; Tomita, M. Grip strength in the frail elderly. *Am. J. Phys. Med. Rehabil.* **2004**, *83*, 819–826. [CrossRef]
9. Lin, C.H.; Chou, L.W.; Wei, S.H.; Lieu, F.K.; Chiang, S.L.; Sung, W.H. Influence of aging on bimanual coordination control. *Exp. Gerontol.* **2014**, *53*, 40–47. [CrossRef]
10. Lee, S.C.; Wu, L.C.; Chiang, S.L.; Lu, L.H.; Chen, C.Y.; Lin, C.H.; Ni, C.H.; Lin, C.H. Validating the Capability for Measuring Age-Related Changes in Grip-Force Strength Using a Digital Hand-Held Dynamometer in Healthy Young and Elderly Adults. *BioMed Res. Int.* **2020**, *2020*, 6936879. [CrossRef]
11. Lin, C.H.; Sung, W.H.; Chiang, S.L.; Lee, S.C.; Lu, L.H.; Wang, P.C.; Wang, X.M. Influence of aging and visual feedback on the stability of hand grip control in elderly adults. *Exp. Gerontol.* **2019**, *119*, 74–81. [CrossRef]
12. Kao, C.-H.; Chiang, S.-L.; Chou, L.-W.; Lin, C.-H.; Lu, Y.-H.; Lu, L.-H.; Wang, X.-M.; Lin, C.-H. Validation of Vibration Exercises on Enhancing Muscle Strength and Upper Limb Functionality among Pre-Frail Community-Dwelling Older Adults. *Int. J. Environ. Res. Public Health* **2022**, *19*, 14509. [CrossRef] [PubMed]
13. Lin, B.-S.; Kuo, S.-F.; Lee, I.; Lu, L.-H.; Chen, P.-Y.; Wang, P.-C.; Lai, C.-H.; Wang, X.-M.; Lin, C.-H. The impact of aging and reaching movements on grip stability control during manual precision tasks. *BMC Geriatr.* **2021**, *21*, 703. [CrossRef] [PubMed]
14. Greenlund, K.J.; Keenan, N.L.; Clayton, P.F.; Pandey, D.K.; Hong, Y. Public Health Options for Improving Cardiovascular Health Among Older Americans. *Am. J. Public Health* **2012**, *102*, 1498–1507. [CrossRef] [PubMed]
15. Hewett, T.E.; Ford, K.R.; Levine, P.; Page, S.J. Reaching kinematics to measure motor changes after mental practice in stroke. *Top. Stroke Rehabil.* **2007**, *14*, 23–29. [CrossRef] [PubMed]
16. Page, S.J.; Levine, P.; Leonard, A.; Szaflarski, J.P.; Kissela, B.M. Modified constraint-induced therapy in chronic stroke: Results of a single-blinded randomized controlled trial. *Phys. Ther.* **2008**, *88*, 333–340. [CrossRef]
17. Cirstea, M.C.; Levin, M.F. Improvement of arm movement patterns and endpoint control depends on type of feedback during practice in stroke survivors. *Neurorehabil. Neural Repair* **2007**, *21*, 398–411. [CrossRef]
18. Lai, C.H.; Sung, W.H.; Chiang, S.L.; Lu, L.H.; Lin, C.H.; Tung, Y.C.; Lin, C.H. Bimanual coordination deficits in hands following stroke and their relationship with motor and functional performance. *J. Neuroeng. Rehabil.* **2019**, *16*, 101. [CrossRef]
19. Hendricks, H.T.; van Limbeek, J.; Geurts, A.C.; Zwarts, M.J. Motor recovery after stroke: A systematic review of the literature. *Arch. Phys. Med. Rehabil.* **2002**, *83*, 1629–1637. [CrossRef]
20. Bogaerts, A.C.; Delecluse, C.; Claessens, A.L.; Troosters, T.; Boonen, S.; Verschueren, S.M. Effects of whole body vibration training on cardiorespiratory fitness and muscle strength in older individuals (a 1-year randomised controlled trial). *Age Ageing* **2009**, *38*, 448–454. [CrossRef]
21. Pang, M.Y.; Lau, R.W.; Yip, S.P. The effects of whole-body vibration therapy on bone turnover, muscle strength, motor function, and spasticity in chronic stroke: A randomized controlled trial. *Eur. J. Phys. Rehabil. Med.* **2013**, *49*, 439–450.
22. Morel, D.S.; Marín, P.J.; Moreira-Marconi, E.; Dionello, C.F.; Bernardo-Filho, M. Can Whole-Body Vibration Exercises in Different Positions Change Muscular Activity of Upper Limbs? A Randomized Trial. *Dose-Response* **2018**, *16*, 1559325818804361. [CrossRef] [PubMed]
23. Stania, M.; Juras, G.; Słomka, K.; Chmielewska, D.; Król, P. The application of whole-body vibration in physiotherapy—A narrative review. *Acta Physiol. Hung.* **2016**, *103*, 133–145. [CrossRef] [PubMed]
24. Cochrane, D.J. Vibration exercise: The potential benefits. *Int. J. Sports Med.* **2011**, *32*, 75–99. [CrossRef] [PubMed]

25. Alam, M.M.; Khan, A.A.; Farooq, M. Effect of whole-body vibration on neuromuscular performance: A literature review. *Work* **2018**, *59*, 571–583. [CrossRef] [PubMed]
26. Da Silva, U.; Villagra, H.A.; Oliva, L.L.; Marconi, N.F. EMG activity of upper limb on spinal cord injury individuals during whole-body vibration. *Physiol. Int.* **2016**, *103*, 361–367. [CrossRef]
27. Cristino de Souza, A.L.; Mendonca, V.A.; Coelho de Oliveira, A.C.; Ferreira da Fonseca, S.; Mello Santos, L.M.; Cunha Fernandes, J.S.; Leite, H.R.; Luiz de Mendonca Martins, F.; Marcia Dos Santos, J.; de Fatima Silva, A.; et al. Whole body vibration in the static modified push-up position in untrained healthy women stimulates neuromuscular system potentiating increased handgrip myogenic response. *J. Bodyw. Mov. Ther.* **2020**, *24*, 233–238. [CrossRef]
28. Jones, M.T.; Martin, J.R.; Jagim, A.R.; Oliver, J.M. Effect of Direct Whole-Body Vibration on Upper-Body Muscular Power in Recreational, Resistance-Trained Men. *J. Strength Cond. Res.* **2017**, *31*, 1371–1377. [CrossRef]
29. Ashnagar, Z.; Shadmehr, A.; Hadian, M.; Talebian, S.; Jalaei, S. The effects of whole body vibration on EMG activity of the upper extremity muscles in static modified push up position. *J. Back Musculoskelet. Rehabil.* **2016**, *29*, 557–563. [CrossRef]
30. Costa, V.; da Silva, F.F.; de Lima, R.M.; Mezêncio, B.; Ferreira, J.C. Mechanical Vibration Increases EMG Activity But Does Not Affect Strength Resistance Performance. *J. Prof. Exerc. Physiol.* **2019**, *16*, 120–129.
31. Lee, J.S.; Kim, C.Y.; Kim, H.D. Short-Term Effects of Whole-Body Vibration Combined with Task-Related Training on Upper Extremity Function, Spasticity, and Grip Strength in Subjects with Poststroke Hemiplegia: A Pilot Randomized Controlled Trial. *Am. J. Phys. Med. Rehabil.* **2016**, *95*, 608–617. [CrossRef]
32. Ahn, J.Y.; Kim, H.; Park, C.B. Effects of Whole-Body Vibration on Upper Extremity Function and Grip Strength in Patients with Subacute Stroke: A Randomised Single-Blind Controlled Trial. *Occup. Ther. Int.* **2019**, *2019*, 5820952. [CrossRef] [PubMed]
33. Boo, J.A.; Moon, S.H.; Lee, S.M.; Choi, J.H.; Park, S.E. Effect of whole-body vibration exercise in a sitting position prior to therapy on muscle tone and upper extremity function in stroke patients. *J. Phys. Ther. Sci.* **2016**, *28*, 558–562. [CrossRef] [PubMed]
34. Pollock, R.D.; Woledge, R.C.; Mills, K.R.; Martin, F.C.; Newham, D.J. Muscle activity and acceleration during whole body vibration: Effect of frequency and amplitude. *Clin. Biomech.* **2010**, *25*, 840–846. [CrossRef] [PubMed]
35. Cardinale, M.; Wakeling, J. Whole body vibration exercise: Are vibrations good for you? *Br. J. Sports Med.* **2005**, *39*, 585–589. [CrossRef]
36. Lee, G. Does whole-body vibration training in the horizontal direction have effects on motor function and balance of chronic stroke survivors? A preliminary study. *J. Phys. Ther. Sci.* **2015**, *27*, 1133–1136. [CrossRef]
37. Morel, D.S.; Moreira-Marconi, E.; Neto, S.B.S.; Domingos, L.L.P.; de Souza, P.L.; Caputo, D.; Costa, G.D.; de Figueiredo, C.F.; Carmo, R.C.R.; de Paiva, P.C.; et al. Effects of Whole Body Vibration Intervention on Handgrip Strength of Brazilian Healthy Soldiers. *Afr. J. Tradit. Complement. Altern. Med.* **2017**, *14*, 28–32. [CrossRef]
38. Kiiski, J.; Heinonen, A.; Jarvinen, T.L.; Kannus, P.; Sievanen, H. Transmission of vertical whole body vibration to the human body. *J. Bone Miner. Res.* **2008**, *23*, 1318–1325. [CrossRef]
39. Hazell, T.J.; Jakobi, J.M.; Kenno, K.A. The effects of whole-body vibration on upper- and lower-body EMG during static and dynamic contractions. *Appl. Physiol. Nutr. Metab.* **2007**, *32*, 1156–1163. [CrossRef]
40. Kurt, C.; Pekunlu, E. Acute effect of whole body vibration on isometric strength, squat jump, and flexibility in well-trained combat athletes. *Biol. Sport* **2015**, *32*, 115–122. [CrossRef]
41. Cochrane, D.J.; Hawke, E.J. Effects of acute upper-body vibration on strength and power variables in climbers. *J. Strength Cond. Res.* **2007**, *21*, 527–531. [CrossRef]
42. Sanudo, B.; Taiar, R.; Furness, T.; Bernardo-Filho, M. Clinical Approaches of Whole-Body Vibration Exercises in Individuals with Stroke: A Narrative Revision. *Rehabil. Res. Pract.* **2018**, *2018*, 8180901. [CrossRef] [PubMed]
43. Huang, M.; Pang, M.Y.C. Muscle activity and vibration transmissibility during whole-body vibration in chronic stroke. *Scand. J. Med. Sci. Sports* **2019**, *29*, 816–825. [CrossRef]
44. Lu, J.; Xu, G.; Wang, Y. Effects of whole body vibration training on people with chronic stroke: A systematic review and meta-analysis. *Top. Stroke Rehabil.* **2015**, *22*, 161–168. [CrossRef] [PubMed]
45. Huang, M.; Liao, L.R.; Pang, M.Y. Effects of whole body vibration on muscle spasticity for people with central nervous system disorders: A systematic review. *Clin. Rehabil.* **2017**, *31*, 23–33. [CrossRef]
46. Bogaerts, A.; Delecluse, C.; Claessens, A.L.; Coudyzer, W.; Boonen, S.; Verschueren, S.M. Impact of whole-body vibration training versus fitness training on muscle strength and muscle mass in older men: A 1-year randomized controlled trial. *J. Gerontol. A Biol. Sci. Med. Sci.* **2007**, *62*, 630–635. [CrossRef] [PubMed]
47. Bissonnette, D.R.; Weir, P.L.; Leigh, L.; Kenno, K. The Effects of a Whole-Body Advanced Vibration Exercise Program on Flexibility, Balance, and Strength in Seniors. *Phys. Occup. Ther. Geriatr.* **2010**, *28*, 225–234. [CrossRef]
48. Marin, P.J.; Ferrero, C.M.; Menendez, H.; Martin, J.; Herrero, A.J. Effects of whole-body vibration on muscle architecture, muscle strength, and balance in stroke patients: A randomized controlled trial. *Am. J. Phys. Med. Rehabil.* **2013**, *92*, 881–888. [CrossRef]
49. Tihanyi, T.K.; Horvath, M.; Fazekas, G.; Hortobagyi, T.; Tihanyi, J. One session of whole body vibration increases voluntary muscle strength transiently in patients with stroke. *Clin. Rehabil.* **2007**, *21*, 782–793. [CrossRef]
50. Armstrong, W.J.; Nestle, H.N.; Grinnell, D.C.; Cole, L.D.; Van Gilder, E.L.; Warren, G.S.; Capizzi, E.A. The acute effect of whole-body vibration on the hoffmann reflex. *J. Strength Cond. Res.* **2008**, *22*, 471–476. [CrossRef]
51. Sayenko, D.G.; Masani, K.; Alizadeh-Meghrazi, M.; Popovic, M.R.; Craven, B.C. Acute effects of whole body vibration during passive standing on soleus H-reflex in subjects with and without spinal cord injury. *Neurosci. Lett.* **2010**, *482*, 66–70. [CrossRef]

52. Dias, J.A.; Ovando, A.C.; Külkamp, W.; Borges Junior, N.G. Força de preensão palmar: Métodos de avaliação e fatores que influenciam a medida. *Rev. Bras. Cineantropometria Desempenho Hum.* **2010**, *12*, 209–216.
53. Hermens, H.J.; Freriks, B.; Disselhorst-Klug, C.; Rau, G. Development of recommendations for SEMG sensors and sensor placement procedures. *J. Electromyogr. Kinesiol.* **2000**, *10*, 361–374. [CrossRef] [PubMed]
54. Pujari, A.N.; Neilson, R.D.; Aphale, S.S.; Cardinale, M. Upper limb vibration prototype with sports and rehabilitation applications: Development, evaluation and preliminary study. *Healthc. Technol. Lett.* **2017**, *4*, 44–49. [CrossRef] [PubMed]
55. Cheng, H.Y.; Ju, Y.Y.; Chen, C.L.; Chuang, L.L.; Cheng, C.H. Effects of whole body vibration on spasticity and lower extremity function in children with cerebral palsy. *Hum. Mov. Sci.* **2015**, *39*, 65–72. [CrossRef]
56. Lau, R.W.; Yip, S.P.; Pang, M.Y. Whole-body vibration has no effect on neuromotor function and falls in chronic stroke. *Med. Sci. Sports Exerc.* **2012**, *44*, 1409–1418. [CrossRef]
57. Lee, D.K.; Han, J.W. Effects of active vibration exercise using a Flexi-Bar on balance and gait in patients with chronic stroke. *J. Phys. Ther. Sci.* **2018**, *30*, 832–834. [CrossRef]
58. Liao, L.R.; Ng, G.Y.; Jones, A.Y.; Chung, R.C.; Pang, M.Y. Effects of Vibration Intensity, Exercise, and Motor Impairment on Leg Muscle Activity Induced by Whole-Body Vibration in People With Stroke. *Phys Ther.* **2015**, *95*, 1617–1627. [CrossRef]
59. Bovenzi, M. Metrics of whole-body vibration and exposure-response relationship for low back pain in professional drivers: A prospective cohort study. *Int. Arch. Occup. Environ. Health* **2009**, *82*, 893–917. [CrossRef]
60. Cardinale, M.; Lim, J. The acute effects of two different whole body vibration frequencies on vertical jump performance. *Med. Dello Sport* **2003**, *56*, 287–292.
61. Hwang, K.J.; Ryu, Y.U. Whole body vibration may have immediate adverse effects on the postural sway of stroke patients. *J. Phys. Ther. Sci.* **2016**, *28*, 473–477. [CrossRef]
62. Liao, L.R.; Ng, G.Y.; Jones, A.Y.; Huang, M.Z.; Pang, M.Y. Whole-Body Vibration Intensities in Chronic Stroke: A Randomized Controlled Trial. *Med. Sci. Sports Exerc.* **2016**, *48*, 1227–1238. [CrossRef] [PubMed]
63. Nawayseh, N.; Sinan, H.A.; Alteneiji, S.; Hamdan, S. Effect of gender on the biodynamic responses to vibration induced by a whole-body vibration training machine. *Proc. Inst. Mech. Eng. H—J. Eng. Med.* **2019**, *233*, 383–392. [CrossRef] [PubMed]

biology

MDPI

Review

Near-Infrared Transillumination for Macroscopic Functional Imaging of Animal Bodies

Koichi Shimizu [1,2]

1 School of Optoelectronic Engineering, Xidian University, Xi'an 710071, China; k.shimizu15@kurenai.waseda.jp
2 IPS Research Center, Waseda University, Kitakyushu 808-0135, Japan

Simple Summary: Advancements in optical technology have revitalized transillumination imaging for biomedical research. Using near-infrared (NIR) light, we have visualized internal structures in animals, even under conditions of heavy optical scattering. By extending the Beer–Lambert law's applicability through differentiation principles, we can observe real-time physiological changes noninvasively. A hurdle in transillumination is image blurring caused by scattering in body tissues. We have addressed this difficulty by extracting near-axis scattered components from diffuse light, and by employing software techniques such as PSF deconvolution and deep learning. This shift has enabled the clear 3D imaging of internal structures from blurred 2D images. Validation through experimentation with human and animal subjects has underscored the effectiveness of these techniques. Integrating transillumination with modern technology holds great promise for future biomedical applications.

Abstract: The classical transillumination technique has been revitalized through recent advancements in optical technology, enhancing its applicability in the realm of biomedical research. With a new perspective on near-axis scattered light, we have harnessed near-infrared (NIR) light to visualize intricate internal light-absorbing structures within animal bodies. By leveraging the principle of differentiation, we have extended the applicability of the Beer–Lambert law even in cases of scattering-dominant media, such as animal body tissues. This approach facilitates the visualization of dynamic physiological changes occurring within animal bodies, thereby enabling noninvasive, real-time imaging of macroscopic functionality in vivo. An important challenge inherent to transillumination imaging lies in the image blur caused by pronounced light scattering within body tissues. By extracting near-axis scattered components from the predominant diffusely scattered light, we have achieved cross-sectional imaging of animal bodies. Furthermore, we have introduced software-based techniques encompassing deconvolution using the point spread function and the application of deep learning principles to counteract the scattering effect. Finally, transillumination imaging has been elevated from two-dimensional to three-dimensional imaging. The effectiveness and applicability of these proposed techniques have been validated through comprehensive simulations and experiments involving human and animal subjects. As demonstrated through these studies, transillumination imaging coupled with emerging technologies offers a promising avenue for future biomedical applications.

Keywords: biomedical imaging; diffusion; functional imaging; medical imaging; near-infrared light; near-axis scattered light; noninvasive measurement; scattering; transillumination imaging

1. Introduction

Imaging within the realm of biomedical applications can be categorized into two domains based on object size: macroscopic and microscopic imaging [1–5]. The substantive importance of macroscopic imaging has been demonstrated prominently in medical practices, encompassing X-ray imaging, positron emission tomography (PET), magnetic resonance imaging (MRI), and ultrasonic echo imaging [6–12]. Although these modalities offer undeniable utility, they are not devoid of limitations. Even with recent progress in X-ray detection [13,14], the ionizing radiation inherent to X-ray imaging engenders challenges related to repeated exposure. Similarly, the utilization of PET and MRI is impeded by the considerable scale of the necessary apparatus, thereby hindering seamless bedside deployment. The domain of ultrasound imaging presents difficulty involving a tradeoff between spatial resolution and penetration depth in animal bodies. An additional contender for noninvasive macroscopic structural imaging of animal bodies has emerged: optical imaging [15–20].

The methodology of transillumination imaging, wherein light is projected through one side of an animal body to yield an observable image on the opposite side, has been established [21–25]. This approach, which has been demonstrated historically [26], has not been widely used for the macroscopic imaging of internal body structures in medical practice. This decidedly narrow adoption can be attributed, in part, to the absence of suitable instrumentation for its application. For instance, it has been difficult to use high-power light sources because of the heat dissipated in animal bodies. Furthermore, the sensitivity of image sensors proved insufficient to capture the subtle variations present in transillumination images faithfully. Moreover, the quality of transillumination images notably deteriorated when probing deeper anatomical structures. Recently, a technique was reported that can make animal bodies transparent and which can selectively label the target molecules [27–32]. Although this technique is useful for macroscopic organ imaging, it is not applicable to the real-time imaging of live animals.

The impediments outlined above can be surmounted through contemporary optical technologies [33]. The proliferation of high-intensity light sources operating within a narrow wavelength spectrum, along with the emergence of sensitive image-capturing devices, is currently available. Specifically, for the near-infrared (NIR) wavelength range, we possess lasers that exclude heat-producing wavelengths in water, as well as highly sensitive light detectors developed for optical communication purposes [34–38]. Additionally, progress in data processing techniques, notably demonstrated by deep learning, hold the potential to overcome these challenges [39–48].

In the spectrum spanning from ultraviolet to infrared light, NIR light with a 700–1200 nm wavelength exhibits reduced attenuation when passing through animal bodies [15–20]. Leveraging this characteristic along with the background described previously, we have developed various techniques to innovate transillumination imaging. This paper elucidates some of these techniques devised to realize the noninvasive macroscopic and functionally informative imaging of animal bodies within the context of biomedical applications.

2. Two-Dimensional Transillumination Imaging

Figure 1 presents the fundamental principle of transillumination imaging [21–26,49]. We illuminate the animal body from one side and capture the resulting image from the opposite side (Figure 1a). This technique provides a simple and safe tool to investigate the internal structure of an animal body noninvasively. However, for NIR light, an animal body is regarded as a scattering-dominant turbid medium containing an inhomogeneous, light-absorbing internal structure. The scattering coefficient (μ_s) of human body tissue is typically of approximately 10 mm^{-1} in NIR wavelengths [15–20]. This fact implies that the non-scattered component of light decays rapidly within a few millimeters in body tissue. However, when the body is less than a few centimeters thick, the internal light-absorbing structure can be visualized through transillumination imaging [49]. For instance, in human and animal bodies, we can observe a subcutaneous blood vessel network because of the marked differences in the absorption coefficients (μ_a) of hemoglobin and the surrounding tissues [15–20].

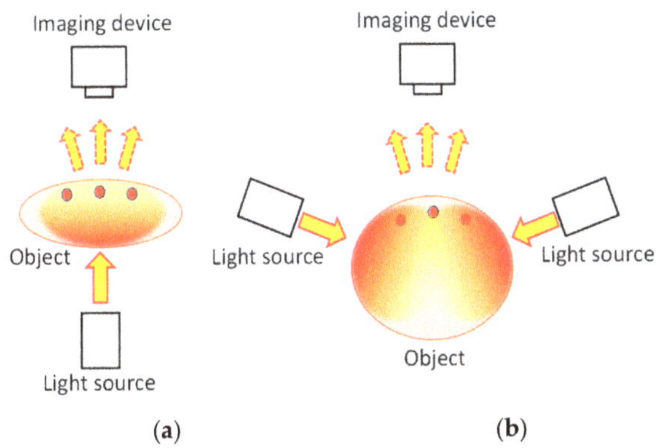

Figure 1. Principle of transillumination imaging: (**a**) transmission mode and (**b**) backscattering mode.

A transillumination image represents the spatial distribution of light that reaches an imaging device through collimation optics, such as a human eye or a camera. In the transmitted light emerging from an animal body, the image-composing light is not the non-scattered component but the component that results from repeated scattering in the forward direction. When we illuminate an animal body with a collimated light beam, we can observe a portion of the light in the same direction as the incident beam even through a few centimeters of thickness. Animal body tissue is known to have an anisotropy factor (g) close to 1 [15–20], which indicates that the scattering pattern of body tissue is highly oriented towards the forward direction. Consequently, some of the light can emerge through the body tissue in the same direction as the incident light after undergoing repeated forward scattering. We refer to this component as near-axis scattered light (NASL) [50]. It shares a similar concept to the snake photon [51–53] or quasi-ballistic light [54–57]. From experimentation, we have confirmed that the NASL intensity remains detectable even after passing through 1 cm thick animal tissue [50]. Because NASL emerges from the animal body in the same direction as the incident light beam, the internal structure can be visualized by employing collimated optics using a regular camera lens with a small aperture.

Figure 2 presents examples of transillumination images of a human and a mouse [49]. Details of the experiments are reported elsewhere [58]. It is noteworthy that this technique can provide real-time video images with simple instrumentation in safe settings, enabling repeated or continuous bedside monitoring. In the video image of the mouse abdomen, the peristaltic movement of its digestive tract during regular food consumption can be visualized, obviating the need for any unnatural contrast medium required for X-ray fluoroscopy.

In cases where the object body is too thick to allow for light transmission in the incident beam direction, indirect light illumination can be used. Figure 1b presents this illumination technique. Given the highly diffusive nature of general body tissue, characterized by a reduced scattering coefficient $\mu_s' = \sim 1\ \mathrm{mm}^{-1}$ in NIR wavelengths, the illuminated light from the side propagates almost uniformly after several millimeters of propagation. Consequently, the illuminated light is dispersed uniformly within the body, serving as back-illumination for imaging subsurface absorbing structures. This illumination technique has found application in finger-vein personal authentication [59,60]. Additionally, its application in the daily management of arterio-venous fistulas in dialysis treatment has been reported [61].

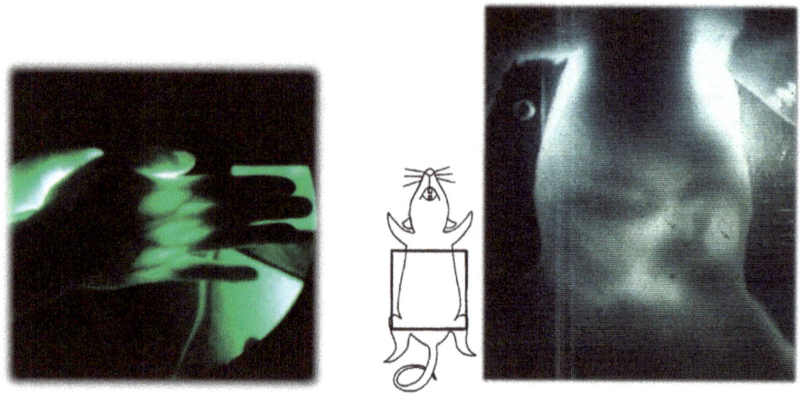

(**a**) (**b**)

Figure 2. Examples of NIR transillumination images: (**a**) human hand and (**b**) mouse abdomen. Modified from Ref. [49].

3. Functional Imaging

One important merit of NIR transillumination imaging is its capability for the non-invasive visualization of physiological changes occurring within animal bodies from an external perspective. Along with the recent progress in optoelectronic devices [62,63], we can leverage the extensive utilization of spectroscopy in the biomedical realm. As explained in detail in Section 2, a transillumination image represents the spatial distribution of NASL observed through collimation optics. The intensity is a consequence of attenuation through animal body tissue, which is a scattering-dominant turbid medium. If we disregard rigor, then the received intensity can be expressed using the following equation.

$$\ln \frac{I_{in}(x,y)}{I_{out}(x,y)} = [\mu_s(x,y) + \mu_a(x,y)]d_t(x,y) - S(x,y) \quad (1)$$

In this equation, I_{in}, I_{out}, μ_{s}, μ_{a}, d_{t}, and S, respectively, denote the incident light intensity, output light intensity (i.e., the transillumination image), scattering coefficient, absorption coefficient, path length in body tissue, and increment factor that represents scattering from off-axis directions. The parameters (x, y) signify the spatial distribution in the plane perpendicular to the optical axis in the z-direction of the Cartesian coordinate system. If we ignore the terms μ_{s} and S, then this equation reduces to the conventional Beer–Lambert law. Because we cannot neglect these terms with animal body tissue, the Beer–Lambert law is typically deemed inapplicable.

However, by obtaining transillumination images under two different conditions, wherein only the absorption condition varies while the other conditions are kept constant, absorption changes in a linear manifestation of the Beer–Lambert law can be visualized as

$$\ln \frac{I_2(x,y)}{I_1(x,y)} = [\mu_{a1}(x,y) - \mu_{a2}(x,y)]d_t(x,y), \quad (2)$$

where designations 1 and 2, respectively, denote conditions before and after the change. Equation (2) elucidates that by calculating the logarithm of intensity ratios for two images, pre-change and post-change, the resulting image delineates the spatial distribution of absorption changes while mitigating the scattering effect. Because μ_{a} is proportional to the molar absorption coefficient ε and the absorber concentration C, the observed absorption changes typically reflect alterations in either ε or C. By capturing transillumination images I_1 and I_2 before and after physiological changes within animal bodies, transillumination imaging can capture the internal functional changes manifesting as variations in ε or C.

One such functional change within animal bodies is the alteration in hemoglobin oxygenation. Figure 3 portrays the absorption spectra of hemoglobin and typical arterial and venous blood, showcasing distinct wavelength-dependent variations in the NIR range. Using these variations, one can visualize the distributions of venous and arterial blood separately. Figure 4 presents the results of venous–arterial differentiation in transillumination images obtained from different human wrists [64].

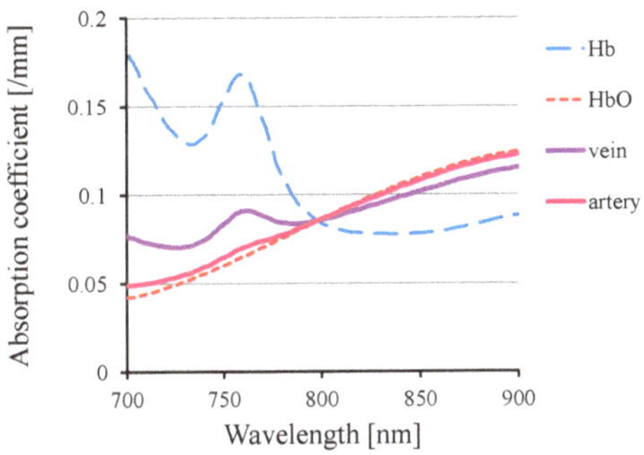

Figure 3. Absorption spectra of hemoglobin and blood.

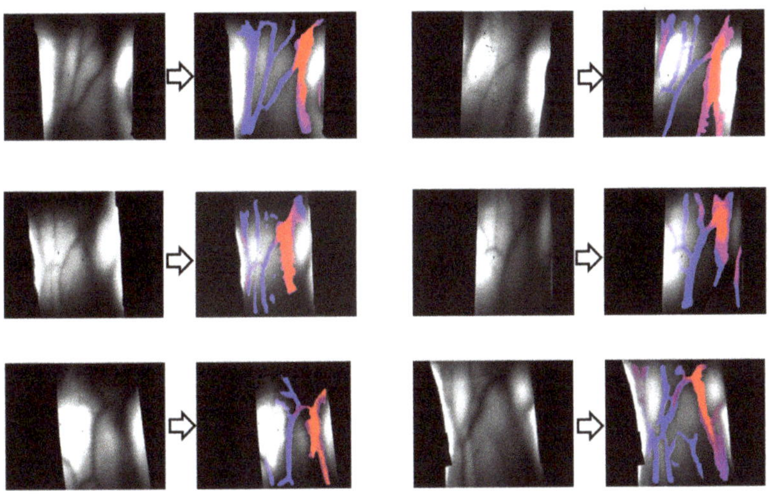

Figure 4. Differentiation of veins (blue color) and arteries (red color) in transillumination images of human adult wrist areas.

In the isosbestic wavelength, oxyhemoglobin and deoxyhemoglobin exhibit equal absorption. Therefore, by using this specific wavelength in the NIR range (typically of approximately 800 nm for human and rodent blood), one can independently visualize alterations in blood volume irrespective of its oxygenation status. Figure 5 displays an example of the functional imaging outcome, achieved through transillumination images captured before and after a physiological change. In this instance, the suppression of renal circulation because of constricted blood vessels was shown for one of the paired mouse kidneys. This result highlighted the alterations in blood volume within the kidney [58,65].

Figure 6 presents the outcomes of functional imaging in a rat brain [65]. Electrostimulation was applied to one of the rat's forelimbs or whisker roots, with the intention of visualizing the corresponding local changes in its brain. Before conducting optical imaging, we identified the cerebral cortex responsible for somatosensory perception. Figure 6a shows the electroencephalograms (EEGs) observed via metal electrodes in direct contact with the mouse head during right forelimb stimulation. Figure 6b presents the functional imaging results, obtained from the transillumination images taken before and after forelimb stimulation. The localized increase in light attenuation, brought about by an augmented blood volume, agreed well with the EEG-identified regions. Stimulation of the right forelimb yielded an activated region in the left cerebral hemisphere. Upon switching the stimulation from the right to the left forelimb, the activated area transitioned symmetrically across the longitudinal cerebral fissure that separates the right and left cerebrum hemispheres. As depicted in Figure 6c, stimulation of the right/left whisker root caused different activated patterns compared to the forelimb stimulation, but remained within the respective left and right hemispheres. These positions and the localized dispersion in the somatosensory area closely correlated with measurements obtained through microelectrodes inserted into the brain [66,67]. It is noteworthy that these observations were noninvasive, requiring no incision or skull opening. The sole intervention involved shaving the hair on the rat's head to enhance the clarity of the transillumination images.

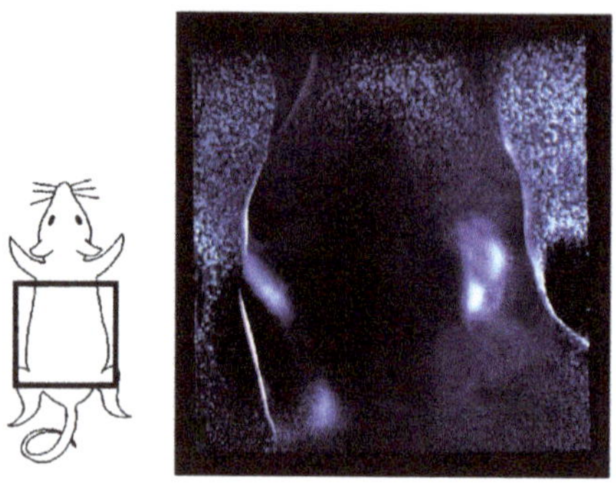

Figure 5. Visualization of blood volume change caused by constricting renal circulation in right kidney. Modified from Ref. [58].

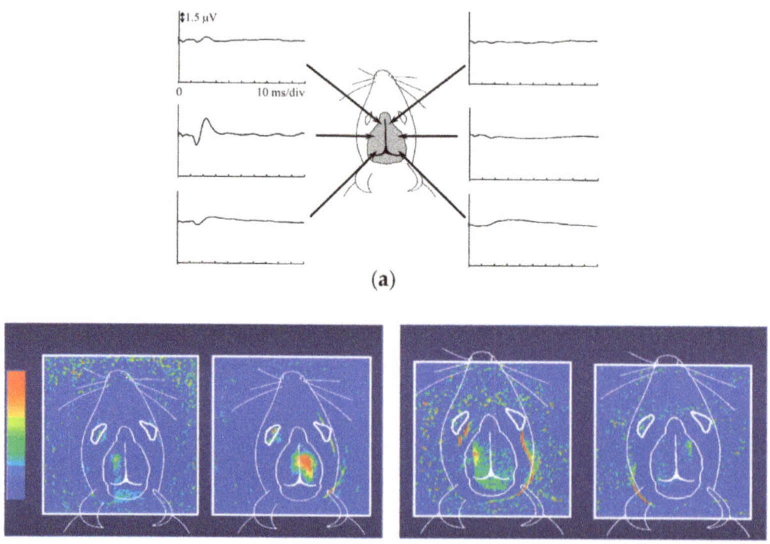

Figure 6. NIR functional transillumination imaging of rat brain: (**a**) EEG identification of activation area for somatosensory stimulation of right forelimb, (**b**) distribution of blood volume change in right/left forearm stimulation, and (**c**) right/left whisker root stimulation.

4. Hardware-Based Scattering Suppression

As elaborated in the preceding chapter, we can improve transillumination imaging by suppressing the strong scattering effects that occur in body tissue. By extracting the NASL component through effective elimination of the dominant diffuse light component, we not only improve transillumination images but also enable computed tomography (CT) using the directly propagated light component [68,69]. Numerous techniques using coherency, polarization, and time-of-flight of the probing light have been reported for extracting the non-scattered component [70–73]. However, because NASL results from repeated forward scattering, these techniques struggle to distinguish NASL from the dominant diffuse light. Therefore, we specifically examined the extraction of the NASL component from the diffuse light. Figure 7 presents details of the principle underpinning our developed NASL extraction method for scattering suppression [50]. Typically, the reduced scattering coefficient of animal tissue ranges around 1 mm^{-1} in NIR wavelength [15–20]. Consequently, the angular distribution of light output from a diffuse medium becomes nearly uniform after traversing body tissue exceeding 1 cm in thickness. Yet, as outlined in Section 2, we anticipate an NASL component output from animal tissue with detectable intensity. Despite its intensity being notably smaller than the dominant diffuse component, we can extract it using the following differential approach [50].

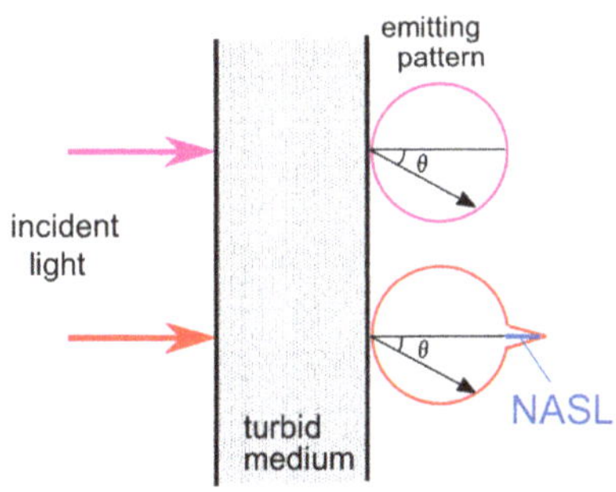

Figure 7. Principle of selective detection of near-axis scattered light (NASL) component from predominant diffuse scattered light: Light distribution of diffuse component only (pink color) and that with NASL component (red color). The NASL component magnitude is exaggerated for illustrative purposes, emphasizing the underlying concept.

When we measure output light from the body surface using a finely collimated detector, the angular distribution of the diffused component is articulated as

$$I_{out}(\theta) = C_o\, I_{in} \cos\theta, \tag{3}$$

where I_{out}, I_{in}, θ ($\neq 0$), and C_o, respectively, stand for the output and incident light intensity, emission angle, and a proportional constant. In the on-axis direction of the incident collimated light beam, the received intensity is as follows.

$$I_{out}(0) = C_o\, I_{in} + I_{NASL} \tag{4}$$

Here, I_{NASL} denotes the NASL intensity. Consequently, we can extract the NASL component by measuring the light intensity at two angles, $\theta = 0$ and θ_m, using the equation

$$I_{NASL} = I_{out}(0) - I_{out}(\theta_m)/\cos\theta_m, \tag{5}$$

where the measurement angle $\theta_m \neq 0$.

To implement this principle, we designed a light detector, as depicted in Figure 8 [50]. To mitigate the diffuse reflection on the body surface, optical fibers in contact with the body surface were employed. The fiber holder on the body surface secured on-axis and off-axis fibers at angles of 0 and 40 degrees, respectively. The outcomes of CT imaging on a mouse abdomen using the developed detector are presented in Figure 9. Cross-sectional images obtained at different heights along the mouse body axis successfully revealed blood-rich organs such as the liver and kidneys [50]. We have developed other hardware-based techniques to suppress scattering effects, and have subsequently confirmed their efficacy through experimentation [74,75].

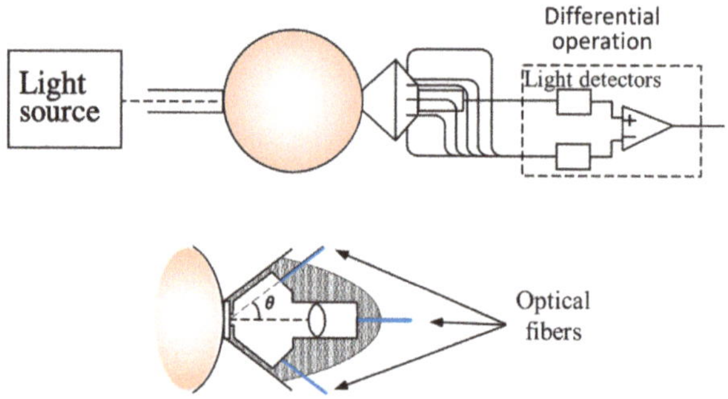

Figure 8. Outline of detector system used to measure NASL intensity from on-axis and off-axis scattered light through the animal body. Modified from Ref. [50].

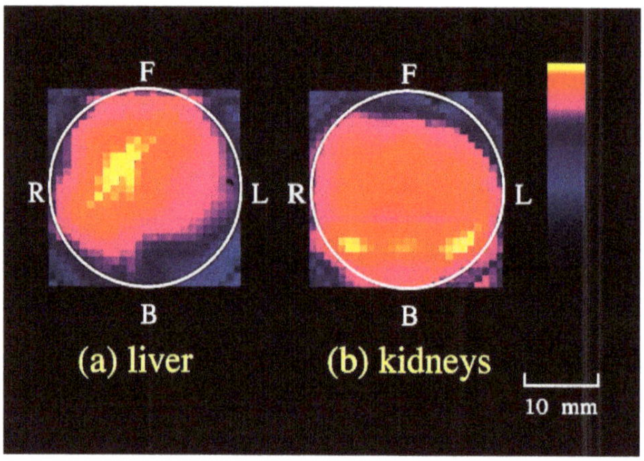

Figure 9. Cross-sectional images of mouse abdomen reconstructed from NASL components. F, B, R and L respectively represent front, back, right and left around the body axis. Modified from Ref. [50].

5. Software-Based Scattering Suppression

The blurring of transillumination images can be modeled as the convolution of the blur-less image with a point spread function (PSF) of the blur. Consequently, if we have a PSF, we can restore the blur-less image by deconvolving the blurred image with the PSF. By applying diffusion approximation to the radiative transfer equation, we derived the theoretical PSF for the model shown in Figure 10 [76].

$$PDF(\rho) = C\left[\left({\mu_s}' + \mu_a\right) + \left(\kappa_d + \frac{1}{\sqrt{\rho^2 + d^2}}\right)\frac{d}{\sqrt{\rho^2 + d^2}}\right]\frac{\exp\left(-\kappa_d\sqrt{\rho^2 + d^2}\right)}{\sqrt{\rho^2 + d^2}} \quad (6)$$

Therein, $\kappa_d = [3\mu_a({\mu_s}' + \mu_a)]^{1/2}$; the radial distance $\rho = (x^2 + y^2)^{1/2}$. In this equation, C, ${\mu_s}'$, μ_a, and d, respectively, represent the constant with respect to ρ and d, the reduced scattering coefficient, the absorption coefficient, and the absorber depth.

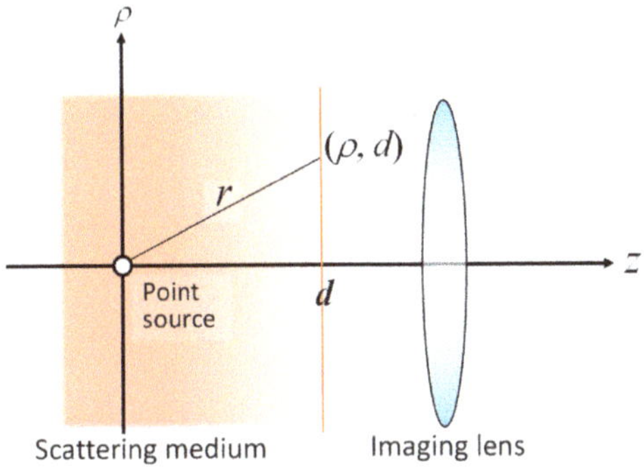

Figure 10. Geometrical model of point-spread function (PSF).

The validity and applicability of the derived PSF were confirmed through experimentation. Figures 11 and 12, respectively, present an outline of the imaging experiments and the results obtained before and after deconvolution [76]. Although the PSF was designed initially for a light source in a turbid medium, it can be extended to a transillumination image. A transillumination image is commonly a distribution of light-absorbing structures in the medium. Nevertheless, if the structure is considered as an ensemble of light-absorbing or light-absent points, then one can employ the same equation (Equation (6)) as the PSF for the transillumination image [77].

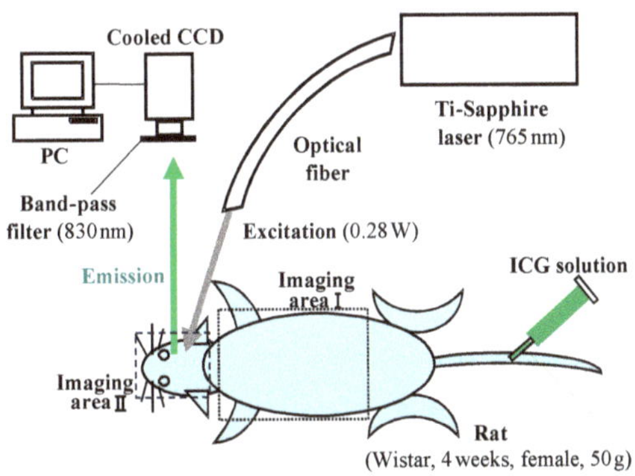

Figure 11. Transcutaneous fluorescent imaging of rat to verify applicability of the PSF deconvolution technique. Modified from Ref. [76].

The PSF of Equation (6) is formulated as a function of the absorber depth d. Consequently, appropriate scattering suppression is applied to a specified depth. Figure 13 portrays the difference between adult arm images acquired through deconvolution with different depth parameters. With small d values in the PSF, shallow veins were deblurred clearly. With large d values in the PSF, deep veins and arteries were deblurred appropriately [78].

Another software-based scattering suppression method uses the deep learning principle. When observing a blurred image through a turbid medium, the observer's mind effectively envisions the unblurred image as if the medium were clear. This cognitive process, which has typically been honed through experience or training, can be implemented within a deep neural network [79]. In a computational framework, a neural network is trained using numerous pairs of training data, consisting of sets of blur-less images and their corresponding blurred counterparts. After extensive training, this network can output blur-less images from previously unseen blurred inputs. Although accumulating a sufficient number of training pairs of transillumination images might prove challenging with specific biomedical applications, the PSF convolution method described above enables the generation of numerous blurred images. We applied this technique to blurred transillumination images of subcutaneous blood vessel networks.

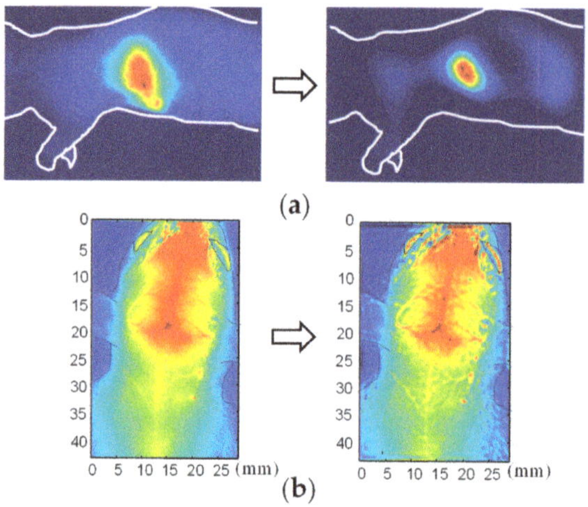

Figure 12. Results of PSF deconvolution: heart (**a**) and fine blood vessels (**b**) became identifiable. Modified from Ref. [76].

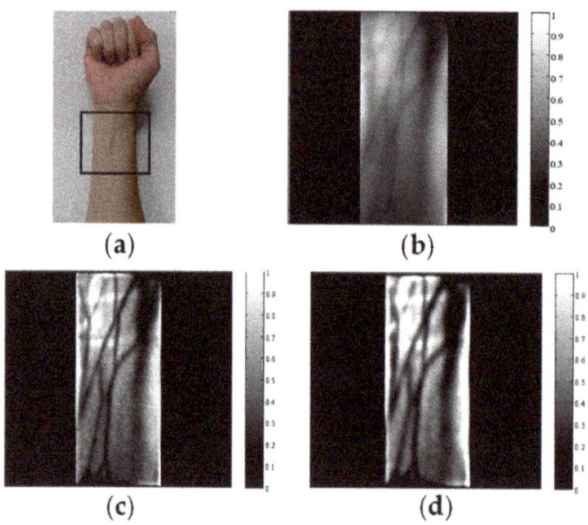

Figure 13. Results of PSF deconvolution: (**a**) visible image and ROI, (**b**) NIR transillumination image, (**c**) deconvolution with the PSF of 2.1 mm depth, and (**d**) deconvolution with the PSF of 4.0 mm depth.

To evaluate the feasibility of this approach, we conducted simulations using blood vessel patterns. For the neural network model, we employed a fully convolutional neural network (FCN) based on a U-net architecture with skip connections [80–83]. Our training dataset comprised blur-less blood vessel images and blurred images generated using the PSF of Equation (6) at varying depths d. By altering the orientation and depth, we generated multiple images from a single original blood vessel image. Figure 14 depicts the dependence of scattering suppression efficacy on the number of training data. The correlation coefficient between the blur-less (ground truth) image and the output image was improved with a sufficient number of training data. Following extensive training with diverse data, the trained FCN effectively outputs clear blur-less images for blurred inputs obtained at different absorber depths, all without requiring depth information.

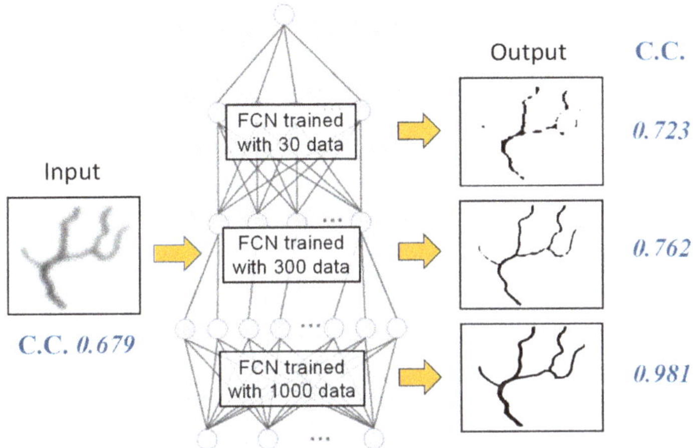

Figure 14. Dependence of output image quality on the number of training data in scattering suppression with FCN. C.C.: correlation coefficient to blur-less (ground truth) image.

6. 3D Transillumination Imaging

In the realm of biomedical applications, the significance of three-dimensional (3D) imaging for assessing the macroscopic internal structure of animal bodies has been exemplified in X-ray CT, MRI, and ultrasound echo imaging. As shown in Section 2, transillumination images inherently remain two-dimensional (2D). During the process of scattering suppression, we recognized that depth information resides within a blurred 2D image. According to the PSF of Equation (6), the extent of blurring and the depth of the absorbing structure share a one-to-one correspondence. In other words, one can deduce the depth from the blurriness present in a single 2D transillumination image.

Building on this concept, we have developed two techniques for reconstructing 3D structures from a single-shot transillumination image [49]. One approach targets curvilinear structures such as subcutaneous blood vessels. The other approach specifically examines solid objects, such as cancer lesions within normal tissue. For the first technique, we established a database that provides absorber depths based on inputs such as the spread and contrast of the blurred image. The transillumination image of the curvilinear structure was partitioned into smaller segments. The depth of each segment was then estimated using the database. The validity of this technique was confirmed through experimentation using an absorber at a known depth in a turbid medium simulating human body tissue. Figure 15 presents an example of depth estimation results. A light-absorbing straight stick was placed at an angle in a tissue-simulating turbid medium, as shown in Figure 15a. Using the proposed technique, we were able to estimate the linear variation of the depth, reaching approximately 25 mm, from a single transillumination image, shown in Figure 15b [49].

The second technique leverages the focus-stacking principle [84–88]. Using the deconvolution operation on a blurred transillumination image with a depth-dependent PSF (such as Equation (6)), only the image portion at a specific depth d is focused correctly. Through iterative deconvolution at varying depths, we acquire multiple images with different in-focus regions of known depths. Using these images, one can compile a composite image in which all segments are in focus and all are annotated with their respective depths. By concurrently applying scattering suppression and depth mapping, it is possible to reconstruct a deblurred 3D image of absorbing structures in a turbid medium. The effectiveness of this technique was examined through experimentation. Figure 16 depicts the object's appearance, its transillumination image in a tissue-simulating turbid medium, and the resultant 3D reconstruction. These analyses confirm the feasibility of obtaining 3D internal structures from a single blurred transillumination image [49].

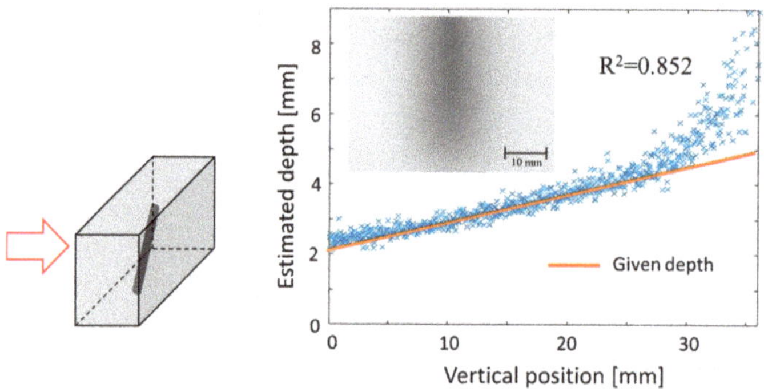

(a) **(b)**

Figure 15. Depth estimation of curvilinear structure in turbid medium: (**a**) light-absorbing bar in tissue-simulating medium, and (**b**) depth estimation from a blurred transillumination image. Modified from Ref. [49].

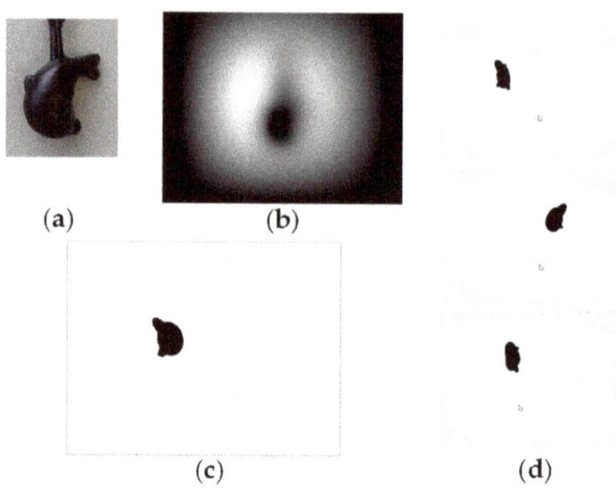

Figure 16. Three-dimensional imaging of a solid object using focus stacking principle: (**a**) object appearance, (**b**) transillumination image in turbid medium, (**c**) result of 3D reconstruction, and (**d**) reconstructed object viewed from different angles. Modified from Ref. [49].

To enhance the utilization of the one-to-one correspondence between blurriness and depth, we incorporated the principles of deep learning into this task. When observing objects at different depths in murky water, the degree of blurring is contingent on the depth. By repeatedly experiencing this phenomenon while having knowledge of the object depths, one can learn to estimate depth from the degree of blur. Therefore, by training a neural network on a computer with numerous training pairs of blurred images and their corresponding depths, a neural network capable of outputting depth for a given blurred image input can be developed. The feasibility of this concept was verified through experimentation. Figure 17 portrays the object structure, its 2D transillumination image through a tissue-simulating medium, and the resultant 3D reconstruction. The deep learning network (FCN) explained in Section 5 eliminated image blur, whereas the 3D structure was reconstructed using a convolutional neural network (CNN) trained on 60,000 pairs of images generated through convolution with the PSF of Equation (6) [79].

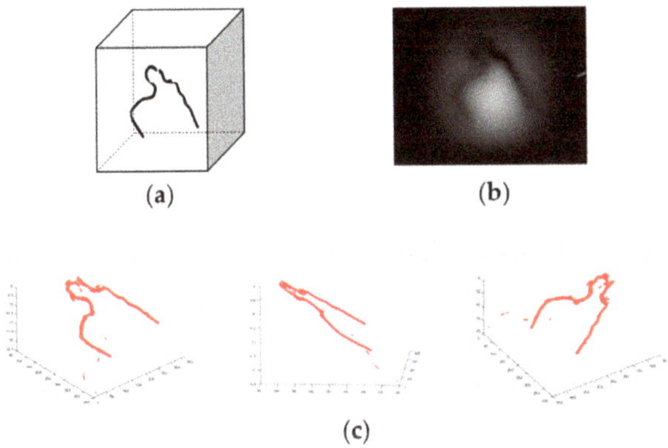

Figure 17. Clear 3D imaging from single transillumination image using FCN and CNN: (**a**) object structure, (**b**) transillumination image in turbid medium, and (**c**) 3D images viewed from different angles. Modified from Ref. [79].

The 3D reconstruction from a single-shot transillumination image proves to be particularly valuable in practical scenarios where imaging opportunities are limited to just one attempt. However, this method entails an important limitation in that it cannot restore structures within completely shadowed areas. For situations in which multiple transillumination imaging from various perspectives is feasible, we have developed an alternative technique. For this approach, scattering effects are mitigated through repeated PSF deconvolution combined with the back-projection algorithm in CT reconstruction [77]. The effectiveness of this technique was demonstrated in animal experiments. Figure 18 outlines the experimental setup, presents an example of a transillumination image, and showcases the resulting 3D reconstruction of the absorbing structure within a mouse abdomen. It is noteworthy that blood-rich organs such as kidneys and portions of the liver were visualized successfully in the 3D image [77].

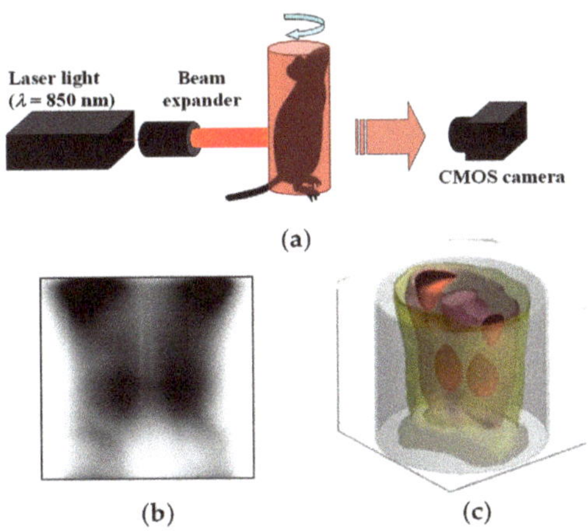

Figure 18. Clear 3D imaging from multiple transillumination images with PSF deconvolution implemented in backpropagation: (**a**) experimental setup, (**b**) transillumination image after deconvolution, and (**c**) reconstructed 3D image of mouse abdomen. Modified from Ref. [77].

7. Conclusions

This review article, written for the Special Issue of "Development of New Optical Techniques and Methods for Basic Biology and Biomedical Applications," introduces an optical technique that has been greatly enhanced by recent technological advances and which is useful today for biomedical applications.

In the pursuit of imaging the internal macroscopic structure of animal bodies, optical transillumination imaging techniques have been innovated through contemporary technologies. Using NIR light, we can effectively capture light that has traversed animal bodies with thicknesses exceeding several cm. Because of strong light scattering in body tissue, non-scattered light passing through an animal body becomes less than the detectable limit for thicknesses of approximately millimeters. However, we noticed the existence of a light-component NASL, which results from repeated forward scattering and which propagates directly through body tissues with thicknesses of approximately centimeters. Through the use of collimation optics, we can acquire transillumination images of internal light-absorbing structures within animal bodies.

This technique enables the visualization of blood vessel networks in human hands, arms, and feet. For thicker body segments, a side illumination method is applicable. In this method, diffused light in the body tissue acts as backlighting, enabling the capture of transillumination images of subsurface structures. This method is useful for the routine management of arteriovenous fistulas, which are constructed in patients' bodies for dialysis treatments. In animal experiments, the movement of internal organs can be monitored through real-time video imaging. The peristaltic motion of the digestive tract during regular food consumption can also be observed without the need for an artificial contrast agent.

Transillumination imaging is also extremely useful for the functional imaging of animal physiology. The intensity ratio between transillumination images before and after a functional change yields an image of the physiological change that occurred within the body. By exploiting differences in the absorption spectra between oxyhemoglobin and deoxyhemoglobin, we have differentiated arteries from veins in human wrists. Using the isosbestic wavelength in the NIR range (approximately 800 nm), it is possible to discern changes in blood volume independently of oxygenation variations. Alterations in renal circulation were visualized in mouse kidney images. The cerebral cortex regions activated via somatosensory stimulation were also visualized in transillumination images of mouse brains. Commissural responses between the right and left hemispheres of the cerebrum were visualized in response to forelimb and whisker stimulation.

A key challenge for biomedical transillumination imaging is image blurring caused by strong light scattering in body tissues. This shortcoming can be addressed by using NASL as the image-forming beam. Based on the differential principle, we have devised a technique to extract the subtle NASL component from the predominant diffuse component in the transmitted light through an animal body. To implement this technique, a detector unit was designed, incorporating optical fiber detectors in both on-axis and off-axis configurations. With this unit integrated into our experimental setup, we achieved optical CT imaging of a mouse abdomen.

The effective suppression of scattering effects can also be accomplished using software-based techniques. A PSF describing the blurring caused by scattering was derived theoretically as a function of the depth of a light-emitting point in a turbid medium. The blurring of transcutaneous fluorescent images was mitigated effectively through PSF deconvolution. This method proved to be applicable to the transillumination imaging of absorbing structures, as well. For instance, the blurring of human blood vessels at varying depths was reduced selectively using the depth-dependent PSF for transillumination images of an adult arm.

An alternative software-based approach involves the application of deep learning principles. An FCN was trained on numerous pairs of blur-less and corresponding blurred images. An extensive set of training pairs was generated by convolving the blur-less image with the derived PSF across various absorber depths. When a new blurred transillumination image was input into the trained FCN, it was able to output a deblurred image, even without prior knowledge of the absorber depth.

Finally, we attempted to elevate the transillumination imaging from 2D to 3D. Subsurface structures of curvilinear absorbing objects in turbid media, such as subcutaneous blood vessels, were reconstructed in 3D using a database capable of providing the absorber's depth for inputting image contrast and blur spread. For solid absorbers, the 3D structure was reconstructed from a single transillumination image using the focus-stacking principle. Using this technique, multiple images were generated by deconvolving the transillumination image using PSFs of varying object depths d. From each segment of the transillumination image, we selected the best focused part among the deconvoluted images, and determined the depth d used for deconvolution. By assembling the chosen segments in their original positions, we were able to reconstruct an all-in-focus image with depth mapping.

Furthermore, the technique with deep learning principles was extended to 3D reconstruction from a single transillumination image. A CNN was trained on pairs of absorber depths and blurred images generated using PSFs of given absorber depths. After sufficient training, the trained CNN was able to output the correct depth of the absorbing structure for the input of a new blurred image. By combining the blur-less image from the FCN and the depth information from the CNN, we reconstructed 3D internal structures in turbid media. If we can capture transillumination images from different perspectives, then 3D reconstruction is attainable without any concern about invisible shadow areas. By incorporating the PSF deconvolution process into the back-projection algorithm, 3D transillumination images of a mouse abdomen were obtained with no shadowed regions.

In conclusion, optical transillumination imaging, although its future potential is recognized, has not yet found widespread practical application. We have continued to innovate this technique using recent technological advancements. Biological safety and practical convenience are essential features of NIR transillumination imaging. Repeated imaging or continuous real-time monitoring using a compact bedside device is now achievable. Moreover, the costs of manufacturing and of the actual use of such equipment are assumed to be much lower than the costs of conventional medical imaging equipment. Noninvasive transillumination imaging of physiological functions will prove to be a valuable tool for various biomedical applications. Furthermore, this advancement will fundamentally reduce the necessity of sacrificing animals for experimentation that conventionally requires open-body observations.

However, the suppression of scattering effects remains a challenge in enhancing the quality of transillumination images. Undoubtedly, continued efforts to address this issue will ultimately engender important advancements and the widespread adoption of the transillumination imaging methodology.

Funding: This study was supported in part by Grants-in-Aid from the Japan Society for the Promotion of Science.

Institutional Review Board Statement: This study was conducted in accordance with the Declaration of Helsinki, and was approved by the Ethics Committee of Hokkaido University (protocol code: 2010-02, date of approval: 28 July 2010). The animal study protocol was approved by the Institutional Review Board of Hokkaido University (protocol code: 08-0127, date of approval: 18 March 2008).

Informed Consent Statement: Informed consent was obtained from all subjects involved in this study.

Data Availability Statement: Data sharing is not applicable to the study described in this article.

Acknowledgments: The author is grateful to Professor Katsuyuki Yamamoto of Hokkaido University for his guidance during the early stages of this study, Professor Yuji Kato of Hokkaido University for his valuable assistance with the experiments, Professor Masataka Kitama of Hokkaido University of Science for his cooperation in CT reconstructions, Professor Trung Nghia Tran of Vietnam National University for his cooperation in animal experiments and Professor Xiaopeng Shao of Xidian University for his support in summarizing this study.

Conflicts of Interest: The author declares that no conflict of interest exist for this report or the study it describes.

References

1. Weissleder, R.; Nahrendorf, M. Advancing biomedical imaging. *Proc. Natl. Acad. Sci. USA* **2015**, *112*, 14424–14428. [CrossRef] [PubMed]
2. Taruttis, A.; van Dam, G.M.; Ntziachristos, V. Mesoscopic and macroscopic optoacoustic imaging of cancer. *Cancer Res.* **2015**, *75*, 1548–1559. [CrossRef] [PubMed]
3. Pian, Q.; Wang, C.; Chen, X.; Liang, J. Multimodal biomedical optical imaging review: Towards comprehensive investigation of biological tissues. *Curr. Mol. Imaging* **2014**, *3*, 72–87. [CrossRef]
4. Ntziachristos, V. Going deeper than microscopy: The optical imaging frontier in biology. *Nat. Methods* **2010**, *7*, 603–614. [CrossRef] [PubMed]
5. Huang, H.K.; Mankovich, N.J.; Altmaier, J. A macroscopic and microscopic biomedical image processing system. *Proc. Annu. Symp. Comput. Appl. Med. Care* **1981**, 580–585.
6. Faragallah, O.S.; El-Hoseny, H.; El-Shafai, W.; El-Rahman, W.A.; El-Sayed, H.S.; El-Rabaie, E.M.; El-Samie, F.E.A.; Geweid, G.G.N. A comprehensive survey analysis for present solutions of medical image fusion and future directions. *IEEE Access* **2021**, *9*, 11358–11371. [CrossRef]
7. Umar, A.A.; Atabo, S.M. A review of imaging techniques in scientific research/clinical diagnosis. *MOJ Anat. Physiol.* **2019**, *6*, 175–183.
8. Bercovich, E.; Javitt, M.C. Medical imaging: From roentgen to the digital revolution, and beyond. *Rambam Maimonides Med. J.* **2018**, *9*, e0034.
9. Kasban, H.; El-Bendary, M.A.M.; Salama, D.H. A comparative study of medical imaging techniques. *Int. J. Inf. Sci. Intell. Syst.* **2015**, *4*, 37–58.
10. Peter, J. Medical Imaging Modalities—An Introduction. In *Advanced Imaging in Biology and Medicine*; Sensen, C.W., Hallgr'ımsson, B., Eds.; Springer: Berlin/Heidelberg, Germany, 2009; Chapter 10; pp. 225–254. [CrossRef]
11. Shung, K.K.; Smith, M.; Tsui, B.M.W. *Principles of Medical Imaging*; Academic Press: San Diego, CA, USA, 2012.
12. Doi, K. Diagnostic imaging over the last 50 years: Research and development in medical imaging science and technology. *Phys. Med. Biol.* **2006**, *51*, R5–R27. [CrossRef]
13. Hellier, K.; Benard, E.; Scott, C.C.; Karim, K.S.; Abbaszadeh, S. Recent progress in the development of a-Se/CMOS sensors for X-ray detection. *Quantum Beam Sci.* **2021**, *5*, 29. [CrossRef]
14. Li, Z.; Zhu, K.; Huang, X.; Zhao, J.; Xu, K. All silicon microdisplay fabricated utilizing 0.18 μm CMOS-IC with monolithic integration. *IEEE Photonics J.* **2022**, *14*, 1–5.
15. Tuchin, V. *Tissue Optics: Light Scattering Methods and Instruments for Medical Diagnosis*, 3rd ed.; The Society of Photo-Optical Instrumentation Engineers (SPIE): Bellingham, WA, USA, 2015.
16. Boas, D.A.; Pitris, C.; Ramanujam, N. *Handbook of Biomedical Optics*; CRC Press: Boca Raton, FL, USA, 2011.
17. Wang, L.V.; Wu, H. *Biomedical Optics: Principles and Imaging*; Wiley: Hoboken, NJ, USA, 2007.
18. Splinter, R.; Hooper, B.A. *An Introduction to Biomedical Optics*; Taylor and Francis: New York, NY, USA, 2007.
19. Vo-Dinh, T. *Biomedical Photonics Handbook*; CRC Press: Boca Raton, FL, USA, 2003.
20. Ntziachristos, V.; Ripoll, J.; Wang, L.V.; Weissleder, R. Looking and listening to light: The evolution of whole-body photonic imaging. *Nat. Biotechnol.* **2005**, *23*, 313–320. [CrossRef]
21. Key, H.; Jackson, P.C.; Wells, P.N.T. New approaches to transillumination imaging. *J. Biomed. Eng.* **1988**, *10*, 113–118. [CrossRef] [PubMed]
22. Gandjbakhche, A.H.; Bonner, R.F.; Nossal, R.; Weiss, G.H. Absorptivity contrast in transillumination imaging of tissue abnormalities. *Appl. Opt.* **1996**, *35*, 1767–1774. [CrossRef] [PubMed]
23. Gurcan, M.N.; Roysam, B.; Dhawan, A.P.; Wang, L.V. Special Letters Issue on Emerging Technologies in Multiparameter Biomedical Optical Imaging and Image Analysis. *IEEE Trans. Biomed. Eng.* **2010**, *57*, 2551–2630. [CrossRef]
24. D'Alessandro, B.; Dhawan, A.P. Depth-dependent hemoglobin analysis from multispectral transillumination images. *IEEE Trans. Biomed. Eng.* **2010**, *57*, 2568–2571. [CrossRef] [PubMed]
25. D'Alessandro, B.; Dhawan, A.P. Transillumination imaging for blood oxygen saturation estimation of skin lesions. *IEEE Trans. Biomed. Eng.* **2012**, *59*, 2660–2667. [CrossRef]
26. Cutler, M. Transillumination of the breast. *Ann. Surg.* **1931**, *93*, 223–234. [CrossRef]
27. Cai, R.; Kolabas, Z.I.; Pan, C.; Mai, H.; Zhao, S.; Kaltenecker, D.; Voigt, F.F.; Molbay, M.; Ohn, T.-L.; Vincke, C.; et al. Whole-mouse clearing and imaging at the cellular level with vDISCO. *Nat. Protoc.* **2023**, *18*, 1197–1242. [CrossRef]
28. Ueda, H.R.; Ertürk, A.; Chung, K.; Gradinaru, V.; Chédotal, A.; Tomancak, P.; Keller, P.J. Tissue clearing and its applications in neuroscience. *Nat. Rev. Neurosci.* **2020**, *21*, 61–79. [CrossRef] [PubMed]

29. Costantini, I.; Cicchi, R.; Silvestri, L.; Vanzi, F.; Pavone, F.S. In-vivo and ex-vivo optical clearing methods for biological tissues. *Biomed. Opt. Express* **2019**, *10*, 5251–5267. [CrossRef]
30. Cai, R.; Pan, C.; Ghasemigharagoz, A.; Todorov, M.I.; Förstera, B.; Zhao, S.; Bhatia, H.S.; Parra-Damas, A.; Mrowka, L.; Theodorou, D.; et al. Panoptic imaging of transparent mice reveals whole-body neuronal projections and skull-meninges connections. *Nat. Neurosci.* **2019**, *22*, 317–327. [CrossRef] [PubMed]
31. Susaki, E.A.; Ueda, H.R. Whole-body and whole-organ clearing and imaging techniques with single-cell resolution: Toward organism-level systems biology in mammals. *Cell Chem. Biol.* **2016**, *23*, 137–157. [CrossRef]
32. Zhu, D.; Larin, K.V.; Luo, Q.; Tuchin, V.V. Recent progress in tissue optical clearing. *Laser Photonics Rev.* **2013**, *7*, 732–757. [CrossRef]
33. Al-Amri, M.D.; El-Gomati, M.; Zubairy, M.S. (Eds.) *Optics in Our Time*; Springer Nature: Cham, Switzerland, 2016.
34. Patle, N.; Raj, A.B.; Joseph, C.; Sharma, N. Review of fibreless optical communication technology: History, evolution, and emerging trends. *J. Optic. Comm.* **2021**. [CrossRef]
35. Sun, X.; Kang, C.H.; Kong, M.; Alkhazragi, O.; Guo, Y.; Ouhssain, M.; Weng, Y.; Jones, H.; Ng, T.K.; Ooi, B.S. A review on practical considerations and solutions in underwater wireless optical communication. *J. Light. Technol.* **2020**, *38*, 421–431. [CrossRef]
36. Kaushal, H.; Kaddoum, G. Optical communication in space: Challenges and mitigation techniques. *IEEE Commun. Surv. Tutor.* **2017**, *19*, 57–96. [CrossRef]
37. Agrawal, G.P. Optical Communication: Its History and Recent Progress. In *Optics in Our Time*; Al-Amri, M.D., El-Gomati, M., Zubairy, M.S., Eds.; Springer Nature: Cham, Switzerland, 2016; pp. 177–199.
38. Hoshi, Y. Functional near-infrared optical imaging: Utility and limitations in human brain mapping. *Psychophysiology* **2003**, *40*, 511–520. [CrossRef]
39. Balasubramaniam, G.M.; Wiesel, B.; Biton, N.; Kumar, R.; Kupferman, J.; Arnon, S. Tutorial on the use of deep learning in diffuse optical tomography. *Electronics* **2022**, *11*, 305. [CrossRef]
40. Zuo, C.; Qian, J.; Feng, S.; Yin, W.; Li, Y.; Fan, P.; Han, J.; Qian, K.; Chen, Q. Deep learning in optical metrology: A review. *Light Sci. Appl.* **2022**, *11*, 39. [CrossRef] [PubMed]
41. Vishnoi, R.; Goswami, P.K. A comprehensive review on deep learning based image dehazing techniques. In Proceedings of the 2022 11th International Conference on System Modeling & Advancement in Research Trends (SMART), Moradabad, India, 16–17 December 2022; pp. 1392–1397. [CrossRef]
42. Hu, K.; Weng, C.; Zhang, Y.; Jin, J.; Xia, Q. An overview of underwater vision enhancement: From traditional methods to recent deep learning. *J. Mar. Sci. Eng.* **2022**, *10*, 241. [CrossRef]
43. Tian, L.; Hunt, B.; Bell, M.A.L.; Yi, J.; Smith, J.T.; Ochoa, M.; Intes, X.; Durr, N.J. Deep learning in biomedical optics. *Lasers Surg. Med.* **2021**, *53*, 748–775. [CrossRef] [PubMed]
44. Gröhl, J.; Schellenberg, M.; Dreher, K.; Maier-Hein, L. Deep learning for biomedical photoacoustic imaging: A review. *Photoacoustics* **2021**, *22*, 100241. [CrossRef] [PubMed]
45. Deng, H.; Qiao, H.; Dai, Q.; Ma, C. Deep learning in photoacoustic imaging: A review. *J. Biomed. Opt.* **2021**, *26*, 040901. [CrossRef] [PubMed]
46. Wang, G.; Ye, J.C.; Man, B.D. Deep learning for tomographic image reconstruction. *Nat. Mach. Intell.* **2020**, *2*, 737–748. [CrossRef]
47. Belthangady, C.; Royer, L.A. Applications, promises, and pitfalls of deep learning for fluorescence image reconstruction. *Nat. Methods* **2019**, *16*, 1215–1225. [CrossRef]
48. Cheng, S.; Li, H.; Luo, Y.; Zheng, Y.; Lai, P. Artificial intelligence-assisted light control and computational imaging through scattering media. *J. Innov. Opt. Health Sci.* **2019**, *12*, 1930006. [CrossRef]
49. Shimizu, K.; Xian, S.; Guo, J. Reconstructing a deblurred 3D structure in a turbid medium from a single blurred 2D image—For near-infrared transillumination imaging of a human body. *Sensors* **2022**, *22*, 5747. [CrossRef]
50. Shimizu, K.; Kitama, M. Fundamental study on near-axis scattered light and its application to optical computed tomography. *Opt. Rev.* **2000**, *7*, 383–388. [CrossRef]
51. Wang, L.; Ho, P.P.; Liu, C.; Zhang, G.; Alfano, R.R. Ballistic 2-D imaging through scattering walls using an ultrafast optical Kerr gate. *Science* **1991**, *253*, 769–771. [CrossRef] [PubMed]
52. Das, B.B.; Yoo, K.M.; Alfano, R.R. Ultrafast time-gated imaging in thick tissues: A step toward optical mammography. *Opt. Lett.* **1993**, *18*, 1092–1094. [CrossRef] [PubMed]
53. Das, B.B.; Yoo, K.M.; Alfano, R.R. Snake Photon Imaging in Thick Tissues in Visible and Near Infrared Radiation Spectral Region. *Adv. Opt. Imaging Photon Migr.* **1994**, *21*, 166–169.
54. Freund, I.; Kaveh, M.; Rosenbluh, M. Dynamic Multiple Scattering: Ballistic Photons and the Breakdown of the Photon-Diffusion Approximation. *Phys. Rev. Lett.* **1988**, *60*, 1130–1133. [CrossRef]
55. Chen, N.G.; Bai, J. Estimation of quasi-straightforward propagating light in tissues. *Phys. Med. Biol.* **1999**, *44*, 1669–1676. [CrossRef]
56. Vasefi, F.; Najiminaini, M.; Ng, E.; Kaminska, B.; Chapman, G.H.; Carson, J.J.L. Angular domain transillumination imaging optimization with an ultrafast gated camera. *J. Biomed. Opt.* **2010**, *15*, 061710. [CrossRef]
57. Gopal, V.; Mujumdar, S.; Ramachandran, H.; Sood, A.K. Imaging in turbid media using quasi-ballistic photons. *Opt. Commun.* **1999**, *170*, 331–345. [CrossRef]

58. Shimizu, K.; Mouri, M.; Yamamoto, K. Trans-body imaging of physiological functions with light. In *Optical Methods in Biomedical and Environmental Sciences*; Ohzu, H., Komatsu, S., Eds.; Elsevier Science: Amsterdam, The Netherlands, 1994; pp. 63–66.
59. Uhl, A.; Busch, C.; Marcel, S.; Veldhuis, R. (Eds.) *Handbook of Vascular Biometrics*; Springer Nature: Cham, Switzerland, 2020.
60. Yang, L.; Yang, G.; Yin, Y.; Zhou, L. A survey of finger vein recognition. In *Biometric Recognition*; Sun, Z., Shan, S., Sang, H., Zhou, J., Wang, Y., Yuan, W., Eds.; Springer: Cham, Switzerland, 2014.
61. Kamiyama, H.; Kitama, M.; Shimizu, H.O.; Yamashita, M.; Kojima, Y.; Shimizu, K. Fundamental study for optical transillumination imaging of arteriovenous fistula. *Adv. Biomed. Eng.* **2021**, *10*, 1–10. [CrossRef]
62. Zhao, M.; Hao, Y.; Zhang, C.; Zhai, R.; Liu, B.; Liu, W.; Wang, C.; Jafri, S.H.M.; Razak, A.; Papadakis, R.; et al. Advances in two-dimensional materials for optoelectronics applications. *Crystals* **2022**, *12*, 1087. [CrossRef]
63. Xu, K. Silicon electro-optic micro-modulator fabricated in standard CMOS technology as components for all silicon monolithic integrated optoelectronic systems. *J. Micromech. Microeng.* **2021**, *31*, 054001. [CrossRef]
64. Matsuda, Y.; Namita, T.; Kato, Y.; Shimizu, K. Arteriovenous discrimination by spectroscopic analysis of transillumination images. *IEICE Tech. Rep.* **2012**, *111*, 151–156.
65. Taka, Y.; Sakatani, K.; Kato, Y.; Shimizu, K. Non—Invasive lmaging of absorption changes in rat brain by NIR transillumination. *Med. Imaging Tech.* **1999**, *7*, 545–555.
66. Tandon, S.; Kambi, N.; Jain, N. Overlapping representations of the neck and whiskers in the rat motor cortex revealed by mapping at different anaesthetic depths. *Eur. J. Neurosci.* **2008**, *27*, 228–237. [CrossRef] [PubMed]
67. Zhang, M.; Watanabe, H.; Sarkisyan, D.; Andersen, M.S.; Nosova, O.; Galatenko, V.; Carvalho, L.; Lukoyanov, N.; Thelin, J.; Schouenborg, J.; et al. Hindlimb motor responses to unilateral brain injury: Spinal cord encoding and left-right asymmetry. *Brain Commun.* **2020**, *2*, fcaa055. [CrossRef]
68. Toida, M.; Kondo, M.; Ichimura, T.; Inaba, H. Two-dimensional coherent detection imaging in multiple scattering media based on the directional resolution capability of the optical heterodyne method. *Appl. Phys. B* **1991**, *52*, 391–394. [CrossRef]
69. Devaraj, B.; Usa, M.; Chan, K.P.; Akatsuka, T.; Inaba, H. Recent advances in coherent detection imaging (CDI) in biomedicine: Laser tomography of human tissues in vivo and in vitro. *IEEE J. Sel. Top. Quantum Electron.* **1996**, *2*, 1008–1016. [CrossRef]
70. Muller, G.; Chance, B.; Alfano, R.R.; Arridge, S.; Beuthan, J.; Gratton, E.; Kaschke, M.; Masters, B.; Svanberg, S.; van der Zee, P. (Eds.) *Medical Optical Tomography: Functional Imaging and Monitoring*; SPIE: Bellingham, WA, USA, 1993.
71. Alfano, R.R. (Ed.) *OSA Proceedings on Advances in Optical Imaging and Photon Migration*; Optical Society of America: Washington, DC, USA, 1994.
72. Alfano, R.R.; Fujimoto, J.G. (Eds.) *OSA Trends in Optics and Photonics on Advantages in Optical Imaging and Photon Migration*; Optical Society of America: Washington, DC, USA, 1996.
73. Chan, K.P.; Devaraj, B.; Yamada, M.; Inaba, H. Coherent detection techniques in optical imaging of tissues. *Phys. Med. Biol.* **1997**, *42*, 855–867. [CrossRef] [PubMed]
74. Takagi, K.; Kato, Y.; Shimizu, K. Extraction of near-axis scattered light for transillumination imaging. *Appl. Opt.* **2009**, *48*, D36–D44. [CrossRef]
75. Takagi, K.; Kakinuma, H.; Kato, Y.; Shimizu, K. CW transillumination imaging by extracting weakly scattered light from strongly diffused light. *Opt. Express* **2009**, *17*, 8332–8342. [CrossRef]
76. Shimizu, K.; Tochio, K.; Kato, Y. Improvement of transcutaneous fluorescent images with a depth-dependent point-spread function. *Appl. Opt.* **2005**, *44*, 2154–2161. [CrossRef]
77. Tran, T.N.; Yamamoto, K.; Namita, T.; Kato, Y.; Shimizu, K. Three-dimensional transillumination image reconstruction for small animal with new scattering suppression technique. *Biomed. Opt. Express* **2014**, *5*, 1321–1335. [CrossRef]
78. Yamamoto, K.; Tran, T.N.; Namita, T.; Kato, Y.; Shimizu, K. Image improvement of absorbing structure in turbid medium for optical transillumination imaging of biological body. *IEICE Tech. Rep.* **2014**, *113*, 115–120.
79. Van, T.N.P.; Tran, T.N.; Inujima, H.; Shimizu, K. Three-dimensional imaging through turbid media using deep learning: NIR transillumination imaging of animal bodies. *Biomed. Opt. Express* **2021**, *12*, 2873–2887. [CrossRef] [PubMed]
80. Long, J.; Shelhamer, E.; Darrell, T. Fully convolutional networks for semantic segmentation. In Proceedings of the IEEE Conference on Computer Vision and Pattern Recognition (CVPR), Boston, MA, USA, 7–12 June 2015; pp. 3431–3440.
81. Djavanshir, G.R.; Chen, X.; Yang, W. A Review of artificial intelligence's neural networks (deep learning) applications in medical diagnosis and prediction. *IT Prof.* **2021**, *23*, 58–62. [CrossRef]
82. Cai, L.; Gao, J.; Zhao, D. A review of the application of deep learning in medical image classification and segmentation. *Ann. Transl. Med.* **2020**, *8*, 713. [CrossRef]
83. Qureshi, I.; Yan, J.; Abbas, Q.; Shaheed, K.; Riaz, A.B.; Wahid, A.; Khan, M.W.J.; Szczuko, P. Medical image segmentation using deep semantic-based methods: A review of techniques, applications and emerging trends. *Inf. Fusion* **2023**, *90*, 316–352. [CrossRef]
84. Jütte, L.; Yang, Z.; Sharma, G.; Roth, B. Focus stacking in non-contact dermoscopy. *Biomed. Phys. Eng. Express* **2022**, *8*, 065022. [CrossRef]
85. Zhang, X. Deep Learning-Based Multi-Focus Image Fusion: A Survey and a Comparative Study. *IEEE Trans. Pattern Anal. Mach. Intell.* **2022**, *44*, 4819–4838. [CrossRef]

86. Brecko, J.; Mathys, A.; Dekoninck, W.; Leponce, M.; VandenSpiegel, D.; Semal, P. Focus stacking: Comparing commercial top-end set-ups with a semi-automatic low budget approach. A possible solution for mass digitization of type specimens. *ZooKeys* **2014**, *464*, 1–23. [CrossRef]
87. Galantucci, L.M.; Lavecchia, F.; Percoco, G. Multistack close range photogrammetry for low cost submillimeter metrology. *J. Comput. Inf. Sci. Eng.* **2013**, *13*, 044501. [CrossRef]
88. Zhang, C.; Bastian, J.; Shen, C.; van den Hengel, A.; Shen, T. Extended depth-of-field via focus stacking and graph cuts. In Proceedings of the 2013 IEEE International Conference on Image Processing, Melbourne, Australia, 15–18 September 2013; pp. 1272–1276.

biology MDPI

Review

Non-Invasive Assessment of Vascular Circulation Based on Flow Mediated Skin Fluorescence (FMSF)

Andrzej Marcinek [1,2,*], Joanna Katarzynska [2], Leslaw Sieron [2], Robert Skokowski [2], Jacek Zielinski [3] and Jerzy Gebicki [1,2,*]

1 Institute of Applied Radiation Chemistry, Lodz University of Technology, 90-924 Lodz, Poland
2 Angionica Ltd., 90-924 Lodz, Poland
3 Department of Athletics, Strength and Conditioning, Poznan University of Physical Education, 61-871 Poznan, Poland
* Correspondence: andrzej.marcinek@p.lodz.pl or andrzej.marcinek@angionica.com.pl (A.M.); jerzy.gebicki@p.lodz.pl or jerzy.gebicki@angionica.com.pl (J.G.); Tel.: +48-42-631-30-96 (A.M.); +48-42-631-31-71 (J.G.)

Simple Summary: The Flow Mediated Skin Fluorescence (FMSF) method, based on the observation of the fluorescence of nicotinamide adenine dinucleotide (NADH) coenzyme from skin tissue, allows for a non-invasive assessment of vascular circulation. The method characterizes dynamic changes in the NADH fluorescence caused by microcirculatory oscillations of endothelial, neurogenic or myogenic origin as well as those observed in response to the post-occlusive reactive hyperemia. It enables assessment of both vasoconstriction and vasodilation. The method defines three major parameters characterizing vascular circulation: reactive hyperemia response, hypoxia sensitivity and normoxia oscillatory index. The FMSF technique seems to be an optimal tool for characterizing macro- and micro-circulation status in a wide range of populations, from healthy physically active people to patients suffering from serious health problems related to vascular dysfunction.

Abstract: Flow Mediated Skin Fluorescence (FMSF) is a new non-invasive method for assessing vascular circulation and/or metabolic regulation. It enables assessment of both vasoconstriction and vasodilation. The method measures stimulation of the circulation in response to post-occlusive reactive hyperemia (PORH). It analyzes the dynamical changes in the emission of NADH fluorescence from skin tissue, providing the information on mitochondrial metabolic status and intracellular oxygen delivery through the circulatory system. Assessment of the vascular state using the FMSF technique is based on three parameters: reactive hyperemia response (RHR), hypoxia sensitivity (HS), and normoxia oscillatory index (NOI). The RHR and HS parameters determine the risk of vascular circulatory disorders and are the main diagnostic parameters. The NOI parameter is an auxiliary parameter for evaluating the state of microcirculation under stress of various origins (e.g., emotional stress, physical exhaustion, or post-infection stress). The clinical data show that the risk of vascular complications is limited among people whose RHR, log(HS), and NOI parameters are not significantly below the mean values determined by the FMSF technique, especially if they simultaneously meet the conditions RHR > 30% and log(HS) > 1.5 (HS > 30), and NOI > 60%.

Keywords: flow mediated skin fluorescence; FMSF technique; NADH fluorescence; microcirculation; flowmotion; reactive hyperemia response; RHR; hypoxia sensitivity; HS; normoxia oscillatory index; NOI

1. Fundamentals of the FMSF Methodology

Lifestyle diseases, including cardiovascular diseases, are the most common cause of death in industrialized countries. Their treatment incurs tremendous cost to the healthcare system. Since current approaches focus mainly on managing symptoms, there is a need for methods and tools for early diagnosis and monitoring of treatment, especially minimally invasive techniques.

There is convincing evidence of the involvement of endothelial dysfunction in the pathogenesis of vascular diseases. Vascular endothelium is an important metabolically active organ that regulates numerous functions and maintains vascular homeostasis, as well as vascular tone and structure—i.e., the dimensions, elasticity, permeability, and reactivity of arteries and veins, including their ability to constrict and dilate. Healthy endothelium is linked to an intact balance of the release of endothelium-derived vasodilating factors (nitric oxide (NO), prostacyclin (PGI2)), endothelium-derived hyperpolarizing factors (EDHFs), and endothelium-derived vasoconstricting factors (endothelin 1, angiotensin II, and thromboxane A2). Under pathological conditions, such as hypercholesterolemia, hypertension, and coronary disease, the production of vasodilating factors is impaired and the production of constricting factors is activated, which also promotes vascular inflammation, remodeling, and thrombosis. A healthy endothelium is essential for undisturbed functioning of the cardiovascular system, while endothelial-dependent dysfunction has been identified as a common thread linking all cardiovascular risk factors. As both an early event and major risk factor, endothelial dysfunction is an important indicator for medical diagnosis. Endothelial dysfunction may be regarded as a barometer of cardiovascular risk [1–4].

Most studies focus on assessment of endothelial function in conduit arteries. However, there is compelling evidence that functional changes in circulation in other organs and the pathogenesis of cardiovascular diseases are correlated with microcirculation dysfunction, which may precede endothelial impairment in large blood vessels [5]. Microcirculation is that part of the cardiovascular system located between the arterial and venous systems, where intensive gas and metabolic exchanges take place. The response of the microvascular vessels to ischemia is crucial to limit the degree of tissue damage. Numerous studies have shown that the skin microcirculation, which is a readily accessible vascular bed, is representative for the assessment of the systemic microcirculation, its dysfunction, and pathologies [6]. It has been shown that the reactivity of the skin microcirculation is disturbed in persons with an increased risk of ischemic heart disease [7]. The cutaneous microcirculation is also considered to be an independent predictor of atherosclerotic damage [8]. The epidermal layer of the skin is not vascularized, so oxygen and nutrients are transported from the dermis by diffusion. This makes epidermal cells a sensitive marker of early disorders of vascular circulation.

Since direct observation of microcirculatory vessels is difficult, microvascular function in the human skin is mainly studied based on skin blood flow, with the use of such techniques as Laser Doppler Flowmetry (LDF) and Laser Speckle Contrast Imaging (LSCI). However, it is possible to also test microvascular blood circulation via changes in skin biochemistry, especially the mitochondrial NADH redox state of the dermal and/or epidermal cells, which depends on blood circulation and is sensitive to its changes. Based on this effect, we have developed a new, non-invasive diagnostic technique called Flow Mediated Skin Fluorescence (FMSF) for assessment of vascular circulation and metabolic regulation. The FMSF technique is based on measurement of nicotinamide adenine dinucleotide (NADH) fluorescence from skin tissue cells. NADH and its oxidized form (NAD^+) play a crucial role in biological systems, as redox coenzymes. The interconversion of the NADH/NAD^+ couple has been studied extensively, including from the mechanistic point of view [9].

As shown in Figure 1, only the reduced NADH form emits significant fluorescence, so the changes in the redox equilibrium of the NADH/NAD^+ couple can be monitored by measurements of skin fluorescence. The fluorescence of NADH is the strongest component in the overall fluorescence emitted from human skin. It is important to emphasize that the penetration of exciting light (340 nm, maximum of NADH absorption band) in skin tissue is low (about 0.3–0.5 mm). A substantial fraction of the exciting light is therefore absorbed by the epidermis and papillary dermis [10–13].

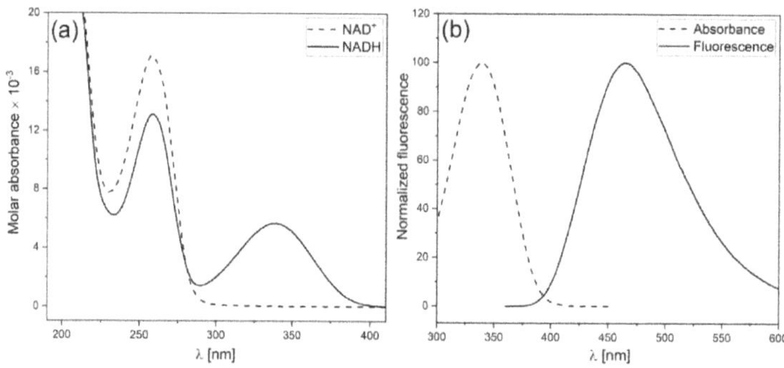

Figure 1. Electronic absorption spectra of NADH and NAD$^+$ (**a**) and fluorescence spectra of NADH (**b**) [14]. Reproduced from Katarzynska, J; Lipinski, Z.; Cholewinski, T., Piotrowski, L; Dworzynski, W.; Urbaniak, M.; Borkowska, A.; Cypryk, K.; Purgal, R.; Marcinek, A.; Gebicki J. Non-invasive evaluation of microcirculation and metabolic regulation using flow mediated skin fluorescence (FMSF): Technical aspects and methodology. *Rev Sci Instrum.* **2019**, *90*, 104104; DOI:10.1063/1.5092218, with the permission of AIP Publishing.

NADH fluorescence has been widely used to determine mitochondrial function in vivo [15,16]. The metabolism of keratinocytes in human skin has also been studied in terms of NADH fluorescence and mitochondrial dynamics [17,18].

As circulation at rest rarely provides significant information on its proper functioning or dysfunction, the FMSF technique, like LDF, uses mechanical stimulation to assess vascular status. It measures changes in the intensity of NADH fluorescence from the skin in response to post-occlusive reactive hyperemia (PORH), which is accomplished by blocking and releasing blood flow in the forearm using a typical occlusion cuff (as used for blood pressure measurement devices). This enables assessment of both vasoconstriction and vasodilation.

PORH is the most popular test used to assess vascular reactivity in both macro and microcirculation, based on the use of mechanical provocation. Studies on the mechanism of action of PORH prove that several important mediators are involved, with nitric oxide (NO) acting as a potent vasodilator of muscle-type arteries [19].

One of the most extensively investigated non-invasive methods using PORH is Flow Mediated Dilation (FMD), which is based on monitoring the diameter of brachial arteries after reactive hyperemia with a two-dimensional ultrasound and Doppler ultrasound [20]. This method has been applied widely in clinics and in large-scale epidemiological research, proving that endothelial dysfunction is associated with cardiovascular diseases. It has been shown in many clinical trials that brachial artery FMD independently predicts the risk of future cardiovascular events and enables monitoring of the effectiveness of treatment [21–28]. Low-Flow Mediated Constriction (L-FMC) can be considered as a complementary method and in combination with FMD gives an integrated score of stimulated vascular function [29–31].

The PORH test accompanied by observation of blood flow changes in the skin microcirculation (by Doppler methods or tonometry, such as Peripheral Arterial Tonometry (PAT)) is also a popular method of assessing endothelial functioning. The blockage of blood flow in main arteries and its subsequent resumption cause quantitative and dynamic changes related to the bioavailability of NO. Laser Doppler Flowmetry assesses blood flow with a high sampling frequency and very good time resolution, enabling analysis of changes in the skin perfusion after stimulation of the microcirculation [6,32]. The time evolution of the post-occlusive signal resembles that observed by FMD. A major disadvantage of LDF is its poor reproducibility, which is caused mainly by the regional heterogeneity of skin perfusion due to skin anatomy and the small size of the monitored region [33].

NADH fluorescence from human skin at rest provides limited information on the level of the NADH coenzyme in skin cells and the redox equilibrium of the NADH/NAD$^+$ pair, as indicators of mitochondrial function. However, the reduced form of the coenzyme (NADH) accumulates under ischemic conditions and during hypoxia, and undergoes oxidation during hyperemia. The degree of change in NADH/NAD$^+$ imbalance as a result of the PORH test can be used to assess the vascular response to ischemic conditions. In the last decade, substantial evidence has linked vascular diseases with a dysfunctional response to hypoxia.

Some similarity in laser Doppler and FMSF post-occlusive signals can be observed, although dynamical changes in NADH fluorescence are slower than the dynamic changes of skin blood flow observed by simultaneous FMSF and LDF measurements. In fact, the measured changes in NADH fluorescence reflect the intrinsic changes of NADH levels in skin tissue, providing important information on the mitochondrial metabolic state in terms of oxygen delivery. Both LDF and PAT measurements remain blind to the ischemic period. In contrast, in the FMSF technique the temporal evolution of the ischemic and hyperemic periods resemble those observed by L-FMC and FMD, respectively. Monitoring the changes in NADH fluorescence from the skin reflecting chances of oxygen and nutrient delivery to epidermal cells by the vascular circulation seems an attractive alternative of the circulatory system monitoring without requiring direct observation of the blood flow itself [13,34].

Observations of blood flow changes in the skin microcirculation or NADH fluorescence from epidermal cells by blocking/restoring the blood flow in the brachial artery raises the question of whether macro- and micro-circulation responses to PORH can be distinguished in the overall response to hypoxia.

Blood flow in microvessels is not a homogeneous process, constant over time between successive heart contractions. To be effective, it must be accompanied by vascular oscillations. Oscillations in the microcirculation, known as flowmotion, are a well-recognized feature of cutaneous blood flow [35,36]. The mechanistic aspects of flowmotion have been the subject of extensive study [37]. The main assessment techniques used are Doppler blood flow tests, using for example a Laser Doppler Flowmeter (LDF), which allow for semi-quantitative characterization of changes in cutaneous blood flow [38,39]. Analysis of the signal recorded with an LDF in the frequency domain makes it possible to extract the components of microcirculation oscillations, which are classified as follows: endogenous, independent of NO (<0.0095 Hz); endogenous NO-dependent (0.0095–0.021 Hz); neurogenic (0.021–0.052 Hz); myogenic (0.052–0.15 Hz); respiratory (0.15–0.62 Hz) and cardiac (0.62–2.00 Hz) [40,41].

Vascular changes are determined by changes in the intracellular calcium level and by the membrane potential of vascular smooth muscle cells. The most important processes related to Ca^{2+} activity include the regulation of muscle contraction and, indirectly, the activation of ion pumps, numerous enzymes, and other target proteins including NO synthase, which determines the release of nitric oxide. Nitric oxide synthase, as a result of increased demand for oxygen, also activates the vascular endothelial growth factor (VEGF) receptor. Hypoxia also activates the hypoxia inducible factor (HIF), which is the stimulus that accelerates the transcription of the VEGF gene. Stabilization of HIF-1α is responsible for the regulation of many physiological processes, including wound healing and adaptation to high-altitudes or physical exercise [42].

Myogenic microcirculatory oscillations, associated with changes in the diameter of microvessels (vasomotion), are a very sensitive measure of the microcirculatory response to hypoxia and can be monitored with high precision using the FMSF technique. Thus, the FMSF technique in conjunction with the use of PORH test allows diagnostics of vascular circulation including not only on dysfunctional blood flow in major arteries, but also microcirculatory response to hypoxia.

2. Technical Aspects of the FMSF Measurement

The FMSF method measures NADH fluorescence from the skin of the forearm in response to blocking and releasing blood flow. Blood flow is blocked in the forearm using a typical occlusion cuff, commonly used to measure blood pressure. Measurements are performed using the AngioExpert, a device constructed by Angionica Ltd. (Lodz, Poland).

The AngioExpert measures the NADH fluorescence (the emitted wavelength of NADH fluorescence is around 460 nm) at a sampling frequency of 25 Hz, excited by ultraviolet radiation with a wavelength of 340 nm [14]. The penetration of exciting light (340 nm) in skin tissue is low (about 0.3–0.5 mm). A substantial fraction of the exciting light is therefore absorbed by the epidermis and papillary dermis [11]. In these skin regions, the density of blood microvessels is low and the changes in NADH fluorescence depend mainly on the supply of oxygen diffused from deeper layers. The inefficient absorption of the excited light by blood components is limited.

For fluorescence measurements, AngioExpert uses a light-emitting diode (LED-UV) with an emission maximum of approx. 340 nm as the NADH excitation light source and a second photodiode as the NADH fluorescence detector at 460 nm. In the optical head, a set of interference filters is used to ensure proper selection of excitation and emitted light. This is an important aspect for the selective observation of the changes in NADH fluorescence, although alone it would not be sufficient to limit the excitation of other endogenous fluorophores present in the skin, such as collagen, tryptophan, elastin, flavin adenine dinucleotide or some glycated products present. However, to a large extent, such limitation is achieved as a result of NADH excitation primarily in the keratinocytes of the epidermal layer. We have shown in our methodological work [14], that upon illumination of the skin with the 340 nm light the fluorescence of NADH dominates in the emission spectrum. What is more, the changes in fluorescence caused by brachial artery occlusion (increase of approx. up to 15–20% of the total emission at 460 nm) can be selectively attributed to the changes in NADH fluorescence, and it is worth noting, that FMSF method is based mainly on the analysis of the changes in fluorescence at 460 nm.

The use of the photodiode technique enables miniaturization of the system and noise reduction. The power of the LED-UV diode does not exceed 8 mW, so the emitted UV radiation does not pose any danger to the skin, even with prolonged exposure. A scheme of the device is shown in Figure 2 [14].

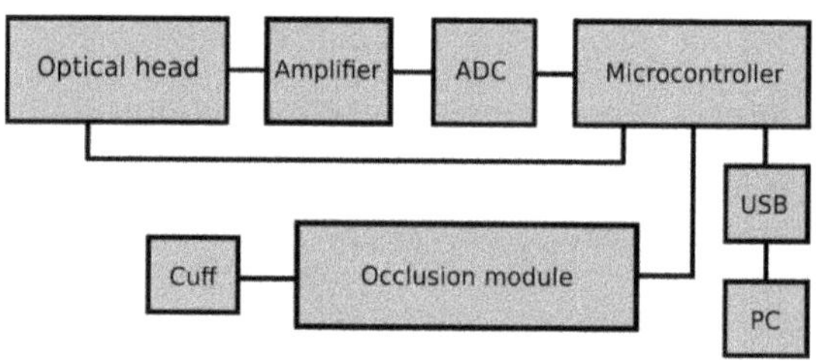

Figure 2. Block scheme of the AngioExpert utilizing FMSF [14]. Reproduced from Katarzynska, J; Lipinski, Z.; Cholewinski, T., Piotrowski, L; Dworzynski, W.; Urbaniak, M.; Borkowska, A.; Cypryk, K.; Purgal, R.; Marcinek, A.; Gebicki J. Non-invasive evaluation of microcirculation and metabolic regulation using flow mediated skin fluorescence (FMSF): Technical aspects and methodology. *Rev Sci. Instrum.* **2019**, *90*, 104104; DOI:10.1063/1.5092218, with the permission of AIP Publishing.

The test is performed with the patient in a comfortable sitting position, after a minimum adaptation period of 5 min, in a quiet room with a controlled air temperature (24 ± 1 °C). The resting NADH fluorescence value emitted by the epidermal layer of the forearm is recorded for the first 3 min (180 s). The brachial artery is then occluded by inflating the cuff of the device to 60 mm Hg above the systolic pressure. The ischemic response is recorded over a period of 3 min (180 s). During this time, ischemic changes in the NADH fluorescence signal are recorded. Upon completion of occlusion, the cuff pressure is released abruptly, restoring flow in the brachial artery and inducing a hyperemic response for a minimum duration of 4 min (240 s).

Although there is no time limit for fluorescence measurements, the time available for measurement and analysis is constrained by the requirement that the patient remain still during the examination, to avoid artifacts and resulting errors. The data are analyzed using analytical software installed on the AngioExpert device or commercially available programs. The initial section of the measurement, during which the patient adjusts to the examination conditions, is discarded from the analysis. Because cuff occlusion is limited to 3 min and the hyperemic/reperfusion period following occlusion also lasts 3 min, analysis of the signal during ischemic and hyperemic periods is limited to that time.

3. Definition of the Measured FMSF Parameters

3.1. Reactive Hyperemia Response (RHR)

Figure 3 presents a typical fluorescence signal for a healthy individual, recorded using the FMSF method by the AngioExpert. During the initial stage of the measurements, the baseline is collected (for 3 min). The FMSF signal is normalized with respect to the mean value of the fluorescence in first 1–2 min of this stage of the measurements. Normalization of the signal makes the result of its analysis independent of measurement conditions related to the individual characteristics of the patient's skin (for example, skin pigmentation, suntan) and/or different technical reasons, as only relative changes are analyzed.

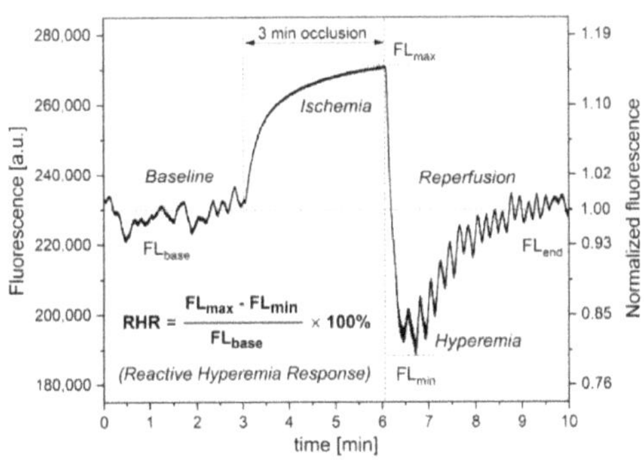

Figure 3. Exemplary FMSF trace recorded for an athlete (male, 40 years); definition of the RHR parameter [43]. Reproduced from: Katarzynska, J.; Zielinski, J.; Marcinek, A.; Gebicki, J. New Approach to Non-Invasive Assessment of Vascular Circulation Based on the Response to Transient Ischemia. *Vasc. Health Risk Manag.* **2022**, *18*, 113–116; DOI:10.2147/VHRM.S358983, Dove Medical Press Ltd. Publisher.

In the second stage of measurements, known as the ischemic response (IR), an increase in NADH fluorescence is observed due to the occlusion of the brachial artery as the cuff is inflated to 60 mm Hg above the systolic blood pressure of the patient. After 3 min, the cuff pressure is released and the NADH fluorescence falls below the baseline, reaching a minimum followed by a return to the baseline. This third measurement stage, called the hyperemic response (HR), consists of a very rapid decrease in NADH fluorescence due to hyperemia (20–30 s) followed by a slow return of NADH fluorescence to baseline due to reperfusion (approximately 3 min).

Based on the combined response from both the ischemic and hyperemic parts of the measured FMSF trace, a Reactive Hyperemia Response (RHR) parameter can be defined [43]. This is a powerful diagnostic tool for characterization of vascular circulation. The RHR parameter characterizes endothelial function related predominantly to the changes in the production of nitric oxide (NO) in the vasculature, mainly in the macrocirculation, due to ischemia and reactive hyperemia ($RHR = IR_{max} + HR_{max}$).

Some additional parameters can be defined that combine the magnitude of changes observed with the rate of NADH fluorescence growth (IR_{index}) during ischemia and the return of fluorescence to the baseline after hyperemia (HR_{index}). In some cases (some patients with type 1 diabetes), the final level of fluorescence does not reach the level of the baseline. This difference is called the Metabolic Recovery (MR) parameter. A detailed definition of these parameters can be found elsewhere [44].

All parameters mentioned above measure the relative changes in NADH fluorescence, expressed as percentages. Using this approach, perturbations caused by the variability of the skin condition are avoided, and comparisons can be made between unhealthy and healthy populations, or between patients.

3.2. Hypoxia Sensitivity (HS)

As can be seen in Figure 3, the FMSF signal oscillates both at rest (basal oscillations called flowmotion at rest (FM)) and even more strongly during the reperfusion period (called flowmotion at the reperfusion period (FM(R)). During the ischemic stage, the signal remains quite smooth because of the blockage of the blood flow, especially in the microcirculation. The altered strength and frequency of oscillations after post-occlusive reactive hyperemia (PORH) reflects the reaction of the vascular microcirculation to hypoxia caused by transient ischemia. Using FMSF signal normalization, two methods can be used to assess the strength of microcirculatory oscillations. The first is based on evaluating the oscillations in terms of the mean square error (MSE), which describes the deviations of the experimental signal points (at a sampling frequency of 25 Hz) from the baseline. In most cases, the baseline around which the FMSF signal oscillates (corresponding to the average fluorescence characteristic for a given patient at rest) is straight. However, in some cases it deviates from a straight horizontal line. Moreover, since during hyperemia the baseline is always an ascending line reaching a plateau the baselines can be defined using the second order polynomial regression method. The mean deviation of the fluorescence signal from the baseline can be used as a measure of the mean magnitude of oscillations. This parameter is objective and patient specific. As MSE values are extremely low, the FM parameters are defined as the MSE values multiplied by a factor of 10^6, to keep them in the number range of units to hundreds. As mentioned, the fluorescence changes are normalized so the FM parameters remain unitless values. The oscillations of the fluorescence signal (seen in Figure 3) relative to baseline, both for the signal before and after occlusion of the brachial artery, are shown in Figure 4a,c.

The second assessment of the strength of oscillations contained in the FMSF signal can be performed using the fast Fourier transform (FFT) algorithm. Power spectral density (PSD) calculated as a mean squared amplitude with rectangular windowing is very well correlated with flowmotion parameters (FM(R)) defined above ($r = 0.996$). Fast Fourier transform analysis provides an estimate of the signal power at a given frequency and its relative contribution to the total power of the signal. The calculated power is grouped into three frequency intervals: $\leq$0.021 Hz, (0.021–0.052 Hz) and (0.052–0.15 Hz). These correspond to endothelial, neurogenic, and myogenic activities, respectively (see Figure 4b).

Among low frequency oscillations, the fraction of FM(R) (or PSD $\times$ 10^6) values covering the intensity of flowmotion related to myogenic oscillations (0.052–0.15 Hz) is especially interesting. Recorded during reperfusion (see Figure 4c,d), this value shows what is called Hypoxia Sensitivity (HS), as it is entirely responsible for the increased activity of the vessels after post-occlusive reactive hyperemia. Thus, the HS parameter, similarly to efficient stabilization of HIF-1α in microvascular smooth muscle cells during transient hypoxia, reflects the microcirculatory response to hypoxia.

Whereas the HS parameter varies within quite a broad range, log(HS) remains normally distributed.

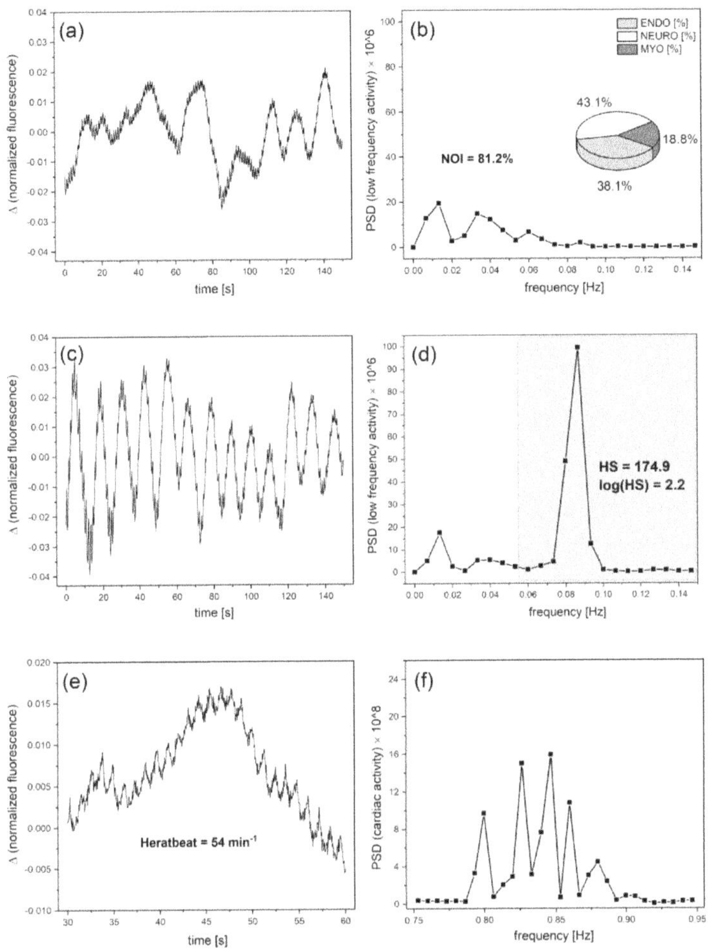

Figure 4. Exemplary oscillations of the FMSF signal relative to baseline with corresponding power spectrum density (PSD) observed for an athlete (male, age 40 years); the pie chart (inset (**b**)) shows the presence of the three components of flowmotion-endothelial (ENDO), neurogenic (NEURO), and myogenic (MYO), respectively: (**a**,**b**) flowmotion at rest; (**b**) definition of NOI [%] = ENDO [%] + NEURO [%] parameter; (**c**,**d**) flowmotion after occlusion during hyperemia and reperfusion; (**d**) definition of HS = PSD(MYO) $\times$ 10^6 parameter; (**e**) heartbeat oscillations seen in the FMSF signal over a 30 s time period; (**f**) cardiac component of the PSD.

3.3. Normoxia Oscillatory Index (NOI)

Although the microcirculation at rest rarely provides significant information about its normal functioning or dysfunction, which requires the use of provocations such as PORH, some exceptions seem to be of particular interest. One such exception may be the analysis of the flowmotion at rest, especially the relative ratio of endothelial and neurogenic oscillations to myogenic oscillations. Thus, a new parameter representing the contribution of endothelial and neurogenic oscillations relative to all oscillations detected at low frequency intervals (<0.15 Hz) can be introduced [45]:

$$\text{NOI} = [\text{PSD(endothelial)} + \text{PSD (neurogenic)}]/[\text{PSD(endothelial)} + \text{PSD (neurogenic)} + \text{PSD(myogenic)}] \times 100\% \quad (1)$$

Despite of the decrease of flowmotion (FM) at rest with age, the NOI parameter remains age-independent. Moreover, in patients with some diseases affecting the vascular system and thus also basal flowmotion, such as diabetes mellitus (group C–diabetes type-2 patients, n = 70 (40 m, 32 f), mean age 63.1 (45–80 y.)), a similar distribution of NOI is observed as for healthy subjects (group B–healthy middle-aged individuals, n = 32 (19 m, 13 f), mean age 38.2 (30–50 y.)) (see Figure 5) [46].

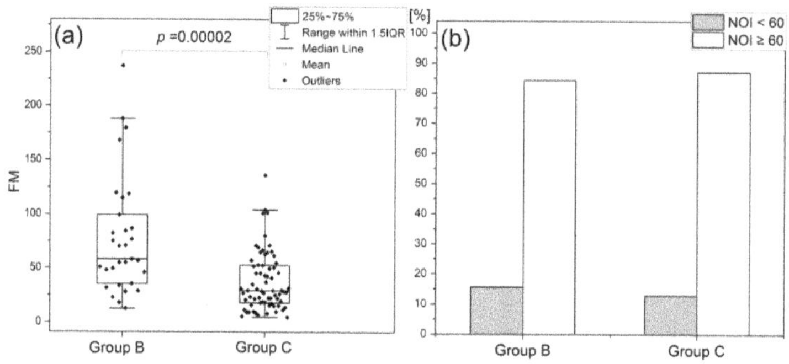

Figure 5. Comparison of the FM (**a**) and NOI (**b**) parameters in groups B–healthy middle-aged individuals, and C–diabetes type-2 patients; the *p*-value was calculated based on the Mann–Whitney test [46]. Based on: Gebicki J, Katarzynska J, Marcinek A. Effect of psychological stress on microcirculation oscillations: diagnostic aspects. *Vasc Health Risk Manag.* **2023**, *19*, 79–82; DOI:10.2147/VHRM.S399082, Dove Medical Press Ltd. publisher.

As has been shown for hundreds of patients investigated using the FMSF method, less than 15% of individuals have an NOI parameter below 60%.

As will be discussed in more detail, there is convincing evidence that a chronic decrease in NOI is associated with various types of stress, such as emotional stress, physical exhaustion, or post-infection stress. It is important to mention that such deviation from the normal NOI distribution may be the result of a significant decrease in endothelial and neurogenic oscillations, with a relatively unchanged value of myogenic oscillations, or, conversely, a significant increase in the myogenic component of basal oscillations at rest.

RHR and log(HS) are the key diagnostic parameters derived from FMSF measurements. They can be used for efficient characterization of vascular circulation based on the response to transient ischemia. NOI is an auxiliary parameter to assess the state of microcirculation under stress of various origins. Chronically low NOI values can lead to the development of serious vascular circulatory disorders.

3.4. Cardiac Oscillations

Heartbeat oscillations (0.6–2 Hz) are also seen in the FMSF signal and PSD, but they remain weak compared to the total PSD and will not be discussed further. It is worth noting, however, that what may appear to be weak signal noise in Figure 4a,c is largely related to cardiac oscillations (see Figure 4e,f).

The weak effect of the heartbeat on changes in total cutaneous NADH fluorescence confirms that a large fraction of the exciting light (340 nm) does not reach the blood vessels directly, as it is mainly absorbed by the epidermis [17]. Skin model Monte Carlo simulations have confirmed that NADH fluorescence is only very weakly affected by blood volume in skin tissue [13].

4. Diagnostic Potential of the FMSF Method

By interpreting the parameters and the dynamics of the NADH fluorescence signal emitted from skin cells in response to occlusion ischemia and subsequent hyperemia, it is possible to identify vascular disorders that can lead to the development of chronic diseases.

4.1. Reactive Hyperemia Response (RHR)

Based on the reactive hyperemia response (RHR parameter) to transient ischemia, the condition of vascular circulation found for three different groups of patients can be distinguished with high statistical significance: A–endurance athletes, n = 50 (33 m, 17 f), mean age 22.0 (16–35 y.); B–healthy middle-aged individuals, n = 32 (19 m, 13 f), mean age 38.2 (30–50 y.); C–diabetes type 2 patients, n = 70 (40 m, 32 f), mean age 63.1 (45–80 y.) (Figure 6). The results presented in Figure 6 b show that individuals with low and high RHR values can be found in each group, but the proportion between the individual RHR ranges is clearly different. In group A, patients with RHR below 25% are the exception, whereas in group C they constitute the dominant majority, while patients with RHR $\geq$ 35% are the exception [43].

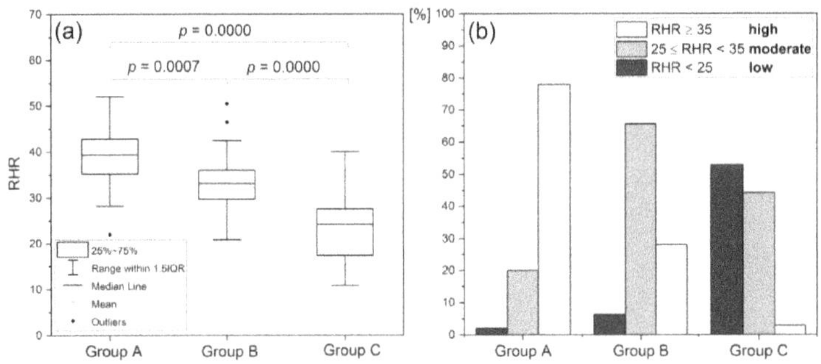

Figure 6. Assessment of RHR parameter in groups A, B, and C: A–highly trained endurance athletes; B–healthy middle-aged individuals; C–diabetes type 2 patients; (**a**) statistical differentiation (one-way ANOVA with the Scheffe post hoc test) of groups A, B, and C based on the RHR parameter; (**b**) distribution of the RHR parameter [43]. Reproduced from: Katarzynska, J.; Zielinski, J.; Marcinek, A.; Gebicki, J. New Approach to Non-Invasive Assessment of Vascular Circulation Based on the Response to Transient Ischemia. *Vasc. Health Risk Manag.* **2022**, *18*, 113–116; DOI:10.2147/VHRM.S358983, Dove Medical Press Ltd. publisher.

Diabetes mellitus (DM) is a chronic metabolic disease that causes vascular complications over time. Excessive production of mitochondrial reactive oxygen species (ROS) is a central mechanism for the development of diabetes complications. It is hypothesized that ROS overproduction may be secondary to impaired responses to hypoxia, due to the inhibition of the hypoxia-inducible factor HIF-1α by hyperglycemia [47,48]. One of the most frequent and severe microvascular complications of DM is diabetic kidney disease (DKD). It has been suggested that vascular endothelial growth factor (VEGF) is critical in glomerular physiology, while DKD is characterized by an imbalance in the expression of VEGF and angiopoietin [49]. HIF-1α is directly responsible for up-regulation of VEGF and is closely related to the etiology of DKD [50,51]. The proper response to hypoxia is one of the mechanisms of kidney adaptation to oxygen deficiency [52].

In research at the Medical University of Lodz (Poland), studies were conducted on a group of 84 volunteers, including 30 patients with diabetic kidney disease (DKD), 33 patients with diabetes mellitus without complications, and 21 healthy control subjects. The results showed that all three groups were clearly differentiated by the RHR parameter with a high statistical significance (RHR(control) = (34.37 $\pm$ 8.18)% vs RHR(DM w/o complications) = (28.75 $\pm$ 7.12)%, $p < 0.05$; RHR(control) = (34.37 $\pm$ 8.18)% vs RHR(DKD) = (18.31 $\pm$ 5.06)%, $p < 0.001$ [53]. The whole studied group showed a significant correlation between the RHR parameter and nephrological parameters, such as serum creatinine concentration (sCr, $r = -0.3$, $p < 0.05$) and estimated glomerular filtration rate (eGFR, $r = -0.61$, $p < 0.001$). Because of their impaired adaptation to hypoxia, patients with diabetes have a higher risk of CVD, and the risk of kidney damage increases by up to 12–17 times [44,54].

4.2. Hypoxia Sensitivity (HS)

A similar conclusion to that formulated on the basis of RHR can be drawn from the analysis of log(HS). The results summarized in Figure 7 were collected for the same three study groups (see Figure 6 for comparison): A–endurance athletes; B–healthy middle-aged individuals; C–diabetes type 2 patients. Again, individuals with low and high log(HS) values can be found in each group, but the proportion between the individual log(HS) ranges is clearly different. Moreover, in group A (highly trained athletes) over 30% of the whole group had HS values above 200 (log(HS) > 2.3). Their microvascular circulation can be described as excellent, and such individuals have strong potential to be outstanding athletes. Similarly, high HS values were hardly ever observed in the other groups. Individuals with very low HS values showed a very weak response to ischemia, which may indicate some microvascular complications. Microvascular complications are unlikely to develop in patients with diabetes type 2 with HS > 100 (log(HS) > 2, less than 10% of the group C population), despite long disease duration and advanced age [55].

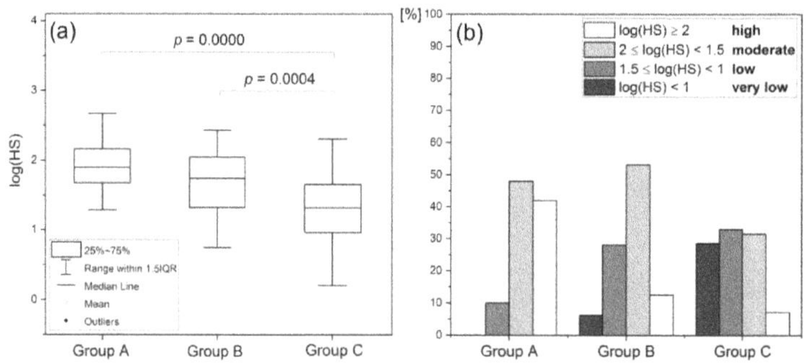

Figure 7. Assessment of log(HS) parameter in various groups: A–highly trained endurance athletes; B–healthy middle-aged individuals; C–diabetes type 2 patients; (**a**) statistical differentiation (one-way ANOVA with the Scheffe post hoc test) of groups A, B, and C based on the log(HS) parameter; (**b**) distribution of the log(HS) parameter [43]. Based on: Katarzynska, J.; Zielinski, J.; Marcinek, A.; Gebicki, J. New Approach to Non-Invasive Assessment of Vascular Circulation Based on the Response to Transient Ischemia. *Vasc. Health Risk Manag.* **2022**, *18*, 113–116; DOI:10.2147/VHRM.S358983, Dove Medical Press Ltd. publisher.

The differences between the groups were partly due to the difference in age, as flowmotion was found to decrease with age [56]. However, this does not affect our general conclusions, especially in terms of individuals with very low log(HS) value finding indicative of microvascular complications. It has also been found that FM(R) (and thus log(HS)) is sensitive to both systolic and diastolic blood pressure, as observed for healthy individuals with normalized blood pressure [56]. This is in agreement with the earlier observation that post-ischemic blood flow oscillations are a sensitive measure of microcirculatory dysfunction in humans with essential arterial hypertension [36].

Similarities in the distribution of log(HS) and RHR parameters among different groups of patients seem to suggest a correlation (for the healthy population $r = 0.413$, $p = 0.0001$) [43]. However, for patients with diseases/disorders of vascular origin, this correlation may be false. For example, the changes in RHR may be greater in the case of CVD, where dysfunction in macrocirculation prevails, regardless of log(HS) values, while the log(HS) parameter can effectively be used to determine the chance of healing in patients with diabetic foot ulcers, regardless of the measured value of the RHR parameter.

In the group of patients with diabetic kidney disease, there is a statistically significant decrease in the log(HS) value found for DKD patients in comparison to healthy subjects (log(HS) = 1.03 vs. log(HS) = 1.59, $p < 0.001$). A significant negative correlation between the log(HS) parameter and sCr and a positive correlation with eGFR were found for the whole studied group (sCr, $r = -0.33$, $p < 0.01$; eGFR, $r = 0.55$, $p < 0.001$) [53].

Diabetic foot ulcers (DFU) are among the most common complications affecting patients with diabetes mellitus. It has been suggested that disturbed hypoxic sensitivity may be responsible for impaired wound healing [47,50,57–60]. Stabilization of HIF-1α may be critical for wound healing, since cellular adaptive responses to hypoxia are mediated by the hypoxia-inducible factor. HIF-1α is also responsible for up-regulation of the expression of vascular endothelial growth factor (VEGF), which is an essential mediator of neovascularization [50]. Since a low log(HS) parameter indicates impaired flowmotion response to hypoxia, it could be also used to predict vascular complications in diabetic foot ulceration.

In a study conducted at the Medical University of Lodz (Poland), 42 patients with DFU and long-term diabetes (over 10 years) were differentiated based on the HS parameter using FMSF. The group of patients with the highest HS values (n = 11, 8m, 3f) did not show any significant disturbances in blood circulation (only one patient had a history of amputation). The second group, composed of patients with the lowest HS values (n = 11, 6m, 5f), showed constant disturbances of vascular circulation. These patients had higher incidences of hypertension, hyperlipidemia, and cardiovascular disease (CVD), as well as neuropathy and nephropathy. Amputations were also more frequent in this group. The HS parameters indicating serious microcirculatory disorders were found to correlate with USG Doppler results indicating macrovascular disturbances ($r = -0.547$, $p = 0.035$) [61]. Since changes in microcirculation occur sooner than in large blood vessels, the log(HS) parameter allows for the early detection of vascular disorders. The FMSF technique can be recommended both for predicting vascular complications in diabetic patients, including DFU, and monitoring their health status.

4.3. Normoxia Oscillatory Index (NOI)

In extensive research conducted at Poznan University of Physical Education (Poland), the FMSF method was used on large groups of professional and amateur athletes. In many cases, hypoxia caused strong activation of myogenic oscillations (see Figure 4), indicating an outstanding predisposition for competitive sports. Moreover, a significant reduction of endothelial and neurogenic oscillations was noticed after exercise to exhaustion (an extreme type of test to assess the athlete's performance) (see Figure 8). High-intensity exercise, accompanied by a corresponding increase in reactive oxygen species (ROS), diminishing NO bioavailability, is known to impose transient stress on the vascular endothelium, as measured by flow-mediated dilatation (FMD) [62,63]. This effect is often called the "exercise paradox." Strenuous physical exercise also affects calcium homeostasis, and this effect persists until the late recovery phase [64,65].

Figure 8 presents the changes in the fluorescence signal relative to the normalized baseline before and after high intensity exercise and the corresponding Power Spectral Density for a highly trained athlete. An evident decrease in the intensity of flowmotion, especially endothelial (<0.021 Hz) and neurogenic (0.021–0.052 Hz) oscillations, can be seen. The flowmotion normalized after 1–3 h of rest and returned to the values observed before exercise. This observation is in accordance with other studies [62,63].

In the aforementioned study on professional and amateur athletes, the strength of endothelial and neurogenic oscillations was measured on the basis of absolute PSD values before and after exercise. However, the flowmotion of a single patient is not always known before factors occur disturbing microcirculatory oscillations. As flowmotion has been found to decrease with age, the reference ranges in different age groups have not yet been established [56]. However, analysis of oscillations based on the NOI parameter (which is independent of age) can also provide reference values. Figure 9 presents the changes in NOI and RHR parameters following strenuous exercise in the group of 26 highly trained amateur runners (n = 26, 26 m, mean age 27.7 (21–40 y.)), who were participating in various competitive sports (long-distance running, cross-country, marathons) [45]. Physical exhaustion can be seen to have a strong effect on both parameters. It should be emphasized that the NOI parameter and the RHR parameter represent distinctive properties of the vascular system. The NOI parameter characterizes microcirculation as it is related to microcirculatory oscillations. In contrast, the RHR parameter characterizes circulation predominantly in large and medium size arteries, as it reflects NO bioavailability caused by reactive hyperemia. It is evident that both parts of the vascular system are affected by strenuous exercise.

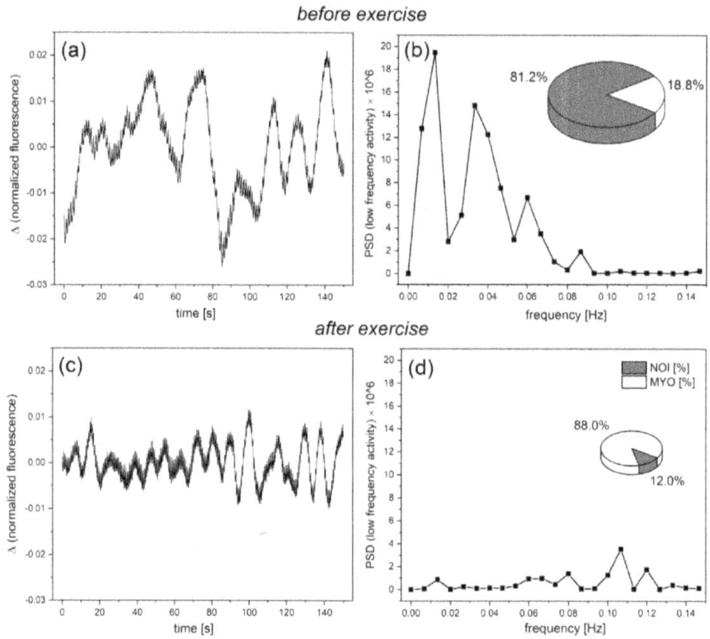

Figure 8. Changes in the oscillations of the FMSF signal recorded for an athlete (male, 40 years) before (**a**) and after (**c**) high-intensity exercise (left) and the corresponding (**b** and **d**) Power Spectral Densities (PSD); the pie charts show the decrease in the NOI parameter [45]. Reproduced from: Chudzik, M.; Cender, A.; Mordaka, R.; Zielinski, J.; Katarzynska, J.; Marcinek, A.; Gebicki, J. Chronic fatigue associated with post-COVID syndrome versus transient fatigue caused by high-intensity exercise: are they comparable in terms of vascular effects? *Vasc Health Risk Manag.* **2022**, *18*, 711–719; DOI:10.2147/VHRM.S371468, Dove Medical Press Ltd. publisher.

Recent years have been dominated by the global health crisis caused by the COVID-19 pandemic and its consequences. The effects of SARS-CoV-2 infection are not limited to the respiratory tract but include serious injuries to vascular circulation, caused by damage to the vascular endothelium [66–68]. Patients recovering from SARS-CoV-2 infection complain of persistent symptoms, the most common of which is fatigue (over 70% of post-COVID patients). There is limited knowledge about the pathophysiology of post-COVID chronic fatigue. Some analogies between post-COVID fatigue syndrome and myalgic encephalomyelitis/chronic fatigue syndrome (ME/CFS) link it to calcium homeostasis dysfunction [69–72]. Since post-COVID syndrome can be linked to permanent vascular dysfunction (as a consequence of the acute phase of COVID-19 infection), it seems rational to compare the chronic vascular effects of post-COVID fatigue with the transient vascular effects of fatigue caused by high-intensity exercise.

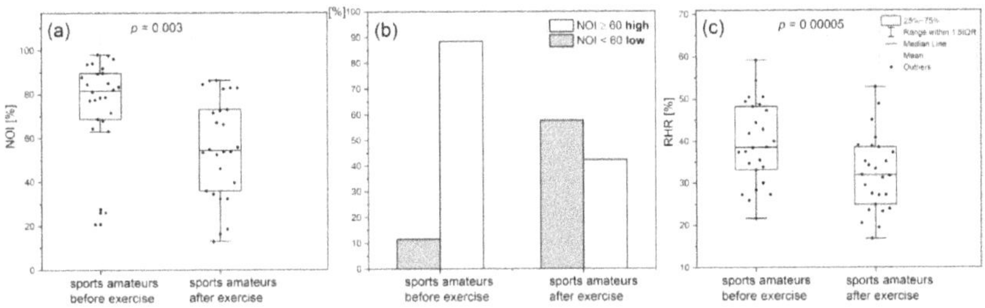

Figure 9. Comparison of NOI (**a**,**b**) and RHR (**c**) parameters in a group of amateur runners before and after strenuous exercise; the *p*-values were calculated by the Wilcoxon signed ranks test (**a**) or the paired sample *t*-test (**c**) [45]. Reproduced from: Chudzik, M.; Cender, A.; Mordaka, R.; Zielinski, J.; Katarzynska, J.; Marcinek, A.; Gebicki, J. Chronic fatigue as-sociated with post-COVID syndrome versus transient fatigue caused by high-intensity exercise: are they com-parable in terms of vascular effects? *Vasc Health Risk Manag.* **2022**, *18*, 711–719; DOI:10.2147/VHRM.S371468, Dove Medical Press Ltd. publisher.

A group of 45 patients (n = 45, 19 m, 26 f, mean age 41.5 (30–50 y.)) with post-COVID syndrome, reporting limited tolerance to exercise and above 50% greater fatigue compared to their pre-COVID levels, was tested using the FMSF technique and compared to a healthy control group (n = 32, 19 m, 13 f, mean age 37.8, (30–50 y.)) [45].

Figure 10 shows a comparison and distribution of the NOI and RHR parameters in the control and post-COVID groups. The vascular effects of post-COVID-related fatigue and exercise-related fatigue measured by the NOI and RHR parameters are quite similar (see Figures 9 and 10). It seems likely that the observed link between the pathophysiology of transient fatigue caused by strenuous exercise and by chronic post-COVID fatigue originates from the modification calcium homeostasis leading to endothelial dysfunction [64,65,69,70].

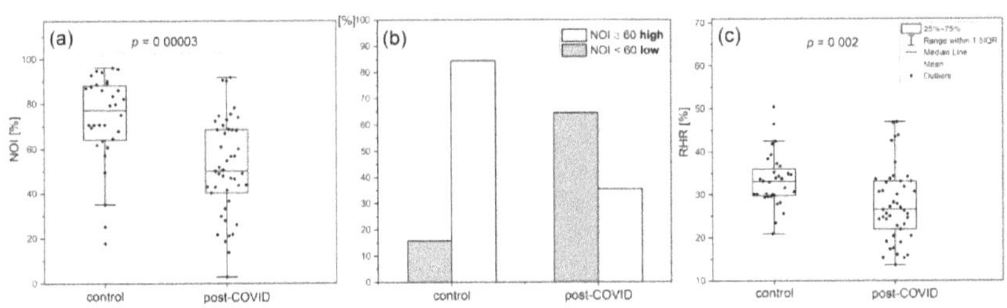

Figure 10. Comparison of NOI (**a**,**b**) and RHR (**c**) parameters in the post-COVID and control group; the *p*-values were calculated by the Mann–Whitney test (**a**) and the two sample *t*-test (**c**) [45]. Reproduced from: Chudzik, M.; Cender, A.; Mordaka, R.; Zielinski, J.; Katarzynska, J.; Marcinek, A.; Gebicki, J. Chronic fatigue as-sociated with post-COVID syndrome versus transient fatigue caused by high-intensity exer-cise: are they com-parable in terms of vascular effects? *Vasc Health Risk Manag.* **2022**, *18*, 711–719; DOI:10.2147/VHRM.S371468, with the permission of Dove Medical Press Ltd.).

The analysis of microcirculation oscillations at rest using the FMSF technique can be of particular use for monitoring physical stress related to strenuous exercise or post-infection fatigue. It is likely that a similar methodology can be applied to the vascular consequences of psychological stress. In numerous individual cases, strong myogenic oscillations in flowmotion at rest have been observed, presumably due to the ischemic response caused by microcirculation vasoconstriction accompanying psychological stress associated with, e.g., traumatic events or cancer disease [46]. In these cases, a strong decrease in the NOI value is also observed, although the changes in the distribution of flowmotion oscillations (activation of the myogenic component) are different. This very attractive potential use of

FMSF technique, to assess the vascular consequences of stress of different origins, will be explored in future work.

5. Concluding Remarks

Flow Mediated Skin Fluorescence (FMSF) is a new non-invasive method enabling assessment of vascular circulation and/or metabolic regulation. The method measures the dynamical changes of NADH fluorescence from the skin tissue, providing information on mitochondrial metabolic status and intracellular oxygen delivery through the circulatory system. It measures stimulation of the circulation in response to post-occlusive reactive hyperemia. It enables assessment of both vasoconstriction and vasodilation.

The FMSF technique is an optimal tool for characterizing macro- and micro-circulation status in a wide range of populations, from healthy physically active people to patients suffering from serious health problems related to vascular dysfunction. Based on three parameters (RHR, log(HS), NOI) measuring the circulatory response to hypoxia, the FMSF technique provides a powerful diagnostic tool for the characterization of vascular circulation.

The diagnostic potential of the RHR parameter includes:

- assessing dysfunction of the vascular circulation, especially the macrocirculation;
- predicting the risk of developing cardiovascular diseases often comorbid with diabetes mellitus.

The diagnostic potential of the log(HS) parameter includes:

- assessing microcirculatory dysfunction in diabetes, cardiovascular disease, peripheral arterial disease, hypertension;
- predicting healing in difficult-to-heal wounds (including diabetic foot ulcers);
- assessing microcirculatory status to determine exercise tolerance in healthy and non-healthy people.

The diagnostic potential of the NOI parameter includes:

- assessing fatigue due to stress from various sources,
- monitoring microcirculation in the recovery and rehabilitation process.

The clinical data show that the risk of vascular complications is limited among people whose RHR, log(HS), and NOI parameters are not significantly below the average values determined by the FMSF technique, especially if they simultaneously meet the condition RHR > 30% and log(HS) > 1.5 (HS > 30), and NOI > 60%.

Author Contributions: A.M.—conceptualization and writing original draft; J.K.—contribution to manuscript preparation and figure visualization; L.S.—contribution to software validation and firmware administration; R.S.—preparation of the computational program and firmware administration; J.Z.—data curation and manuscript review; J.G.—literature search and manuscript review. All authors have read and agreed to the published version of the manuscript.

Funding: This work was supported by the European Union from the resources of the European Regional Development Fund under the Smart Growth Operational Program, Grant No. POIR. 01.01.01-00-0540/15-00.

Institutional Review Board Statement: This is a review article referring to the clinical trial results described in detail in the original articles. Relevant ethics committee approvals are detailed there.

Informed Consent Statement: Not applicable.

Data Availability Statement: Not applicable.

Conflicts of Interest: J.K., L.S., R.S. are employed by Angionica Ltd. J.G. and A.M. are inventors of the patents protecting the use of FMSF technology issued to Angionica Ltd. The authors report no other conflict of interest in this work.

References

1. Vita, J.A.; Keaney, J.F. Endothelial Function. *Circulation* **2002**, *106*, 640–642. [CrossRef] [PubMed]
2. Gryglewski, R.J. Pharmacology of vascular endothelium. *FEBS J.* **2005**, *272*, 2956–2967. [CrossRef] [PubMed]
3. Chłopicki, S.; Gryglewski, R. Angiotensin converting enzyme (ACE) and HydroxyMethylGlutaryl-CoA (HMG-CoA) reductase inhibitors in the forefront of pharmacology of endothelium. *Pharmacol. Rep.* **2005**, *57*, 86–96. [PubMed]
4. Giannotti, G.; Landmesser, U. Endothelial Dysfunction as an Early Sign of Atherosclerosis. *Herz Kardiovaskuläre Erkrankungen.* **2007**, *32*, 568–572. [CrossRef]
5. Hellmann, M.; Roustit, M.; Cracowski, J.L. Skin microvascular endothelial function as a biomarker in cardiovascular diseases? *Pharmacol. Rep.* **2015**, *67*, 803–810. [CrossRef] [PubMed]
6. Cracowski, J.-L.; Roustit, M. Human Skin Microcirculation. In *Comprehensive Physiology*; Prakash, Y.S., Ed.; Wiley: New York, NY, USA, 2020; Chapter 3; Volume 10, pp. 1105–1154. [CrossRef]
7. IJzerman, R.G.; De Jongh, R.T.; Beijk, M.A.M.; Van Weissenbruch, M.M.; Delemarre-van de Waal, H.A.; Serné, E.H.; Stehouwer, C.D.A. Individuals at increased coronary heart disease risk are characterized by an impaired microvascular function in skin. *Eur. J. Clin. Investig.* **2003**, *33*, 536–542. [CrossRef]
8. Rossi, M.; Matteucci, E.; Pesce, M.; Consani, C.; Franzoni, F.; Santoro, G.; Giampietro, O. Peripheral microvascular dysfunction as an independent predictor of atherosclerotic damage in type 1 diabetes patients: A preliminary study. *Clin. Hemorheol. Microcirc.* **2013**, *54*, 381–391. [CrossRef]
9. Gębicki, J.; Marcinek, A.; Zielonka, J. Transient Species in the Stepwise Interconversion of NADH and NAD^+. *Acc. Chem. Res.* **2004**, *37*, 379–386. [CrossRef]
10. McMullen, R.L.; Chen, S.; Moore, D.J. Spectrofluorescence of skin and hair. *Int. J. Cosmet. Sci.* **2012**, *34*, 246–256. [CrossRef]
11. Cicchi, R.; Vogler, N.; Kapsokalyvas, D.; Dietzek, B.; Popp, J.; Pavone, F.S. From molecular structure to tissue architecture: Collagen organization probed by SHG microscopy. *J. Biophotonics* **2013**, *6*, 129–142. [CrossRef]
12. Dunaev, A.V.; Dremin, V.V.; Zherebtsov, E.A.; Rafailov, I.E.; Litvinova, K.S.; Palmer, S.G.; Stewart, N.A.; Sokolovski, S.G.; Rafailov, E.U. Individual variability analysis of fluorescence parameters measured in skin with different levels of nutritive blood flow. *Med. Eng. Phys.* **2015**, *37*, 574–583. [CrossRef] [PubMed]
13. Hou, H.; Du, G.; Wang, Y.; Su, C.; Guo, L.; Chen, X. Noninvasive in vivo study of NADH fluorescence and its real-time intrinsic dynamical changes: Experiments and seven-layered skin model Monte Carlo simulations. *J. Innov. Opt. Health Sci.* **2022**, *15*, 2230006. [CrossRef]
14. Katarzynska, J.; Lipinski, Z.; Cholewinski, T.; Piotrowski, L.; Dworzynski, W.; Urbaniak, M.; Borkowska, A.; Cypryk, K.; Purgal, R.; Marcinek, A.; et al. Non-invasive evaluation of microcirculation and metabolic regulation using flow mediated skin fluorescence (FMSF): Technical aspects and methodology. *Rev. Sci. Instrum.* **2019**, *90*, 104104. [CrossRef]
15. Mayevsky, A.; Rogatsky, G.G. Mitochondrial function in vivo evaluated by NADH fluorescence: From animal models to human studies. *Am. J. Physiol. Physiol.* **2007**, *292*, C615–C640. [CrossRef] [PubMed]
16. Mayevsky, A.; Barbiro-Michaely, E. Use of NADH fluorescence to determine mitochondrial function in vivo. *Int. J. Biochem. Cell Biol.* **2009**, *41*, 1977–1988. [CrossRef]
17. Balu, M.; Mazhar, A.; Hayakawa, C.K.; Mittal, R.; Krasieva, T.B.; König, K.; Venugopalan, V.; Tromberg, B.J. In vivo multiphoton NADH fluorescence reveals depth-dependent keratinocyte metabolism in human skin. *Biophys. J.* **2013**, *104*, 258–267. [CrossRef]
18. Pouli, D.; Balu, M.; Alonzo, C.A.; Liu, Z.; Quinn, K.P.; Rius-Diaz, F.; Harris, R.M.; Kelly, K.M.; Tromberg, B.J.; Georgakoudi, I. Imaging mitochondrial dynamics in human skin reveals depth-dependent hypoxia and malignant potential for diagnosis. *Sci. Transl. Med.* **2016**, *8*, 367ra169. [CrossRef]
19. Roustit, M.; Cracowski, J.L. Assessment of endothelial and neurovascular function in human skin microcirculation. *Trends Pharmacol. Sci.* **2013**, *34*, 373–384. [CrossRef]
20. Celermajer, D.S.; Sorensen, K.E.; Gooch, V.M.; Spiegelhalter, D.J.; Miller, O.I.; Sullivan, I.D.; Lloyd, J.K.; Deanfield, J.E. Non-invasive detection of endothelial dysfunction in children and adults at risk of atherosclerosis. *Lancet* **1992**, *340*, 1111–1115. [CrossRef]
21. Modena, M.G.; Bonetti, L.; Coppi, F.; Bursi, F.; Rossi, R. Prognostic role of reversible endothelial dysfunction in hypertensive postmenopausal women. *J. Am. Coll. Cardiol.* **2002**, *40*, 505–510. [CrossRef]
22. Brevetti, G.; Silvestro, A.; Schiano, V.; Chiariello, M. Endothelial Dysfunction and Cardiovascular Risk Prediction in Peripheral Arterial Disease. *Circulation* **2003**, *108*, 2093–2098. [CrossRef] [PubMed]
23. Gokce, N.; Keaney, J.F.; Hunter, L.M.; Watkins, M.T.; Nedeljkovic, Z.S.; Menzoian, J.O.; Vita, J.A. Predictive value of noninvasively determined endothelial dysfunction for long-term cardiovascular events in patients with peripheral vascular disease. *J. Am. Coll. Cardiol.* **2003**, *41*, 1769–1775. [CrossRef]
24. Chan, S.Y.; Mancini, G.B.J.; Kuramoto, L.; Schulzer, M.; Frohlich, J.; Ignaszewski, A. The prognostic importance of endothelial dysfunction and carotid atheromaburden in patients with coronary artery disease. *J. Am. Coll. Cardiol.* **2003**, *42*, 1037–1043. [CrossRef] [PubMed]
25. Yeboah, J.; Crouse, J.R.; Hsu, F.-C.; Burke, G.L.; Herrington, D.M. Brachial Flow-Mediated Dilation Predicts Incident Cardiovascular Events in Older Adults. *Circulation* **2007**, *115*, 2390–2397. [CrossRef] [PubMed]

26. Huang, A.L.; Silver, A.E.; Shvenke, E.; Schopfer, D.W.; Jahangir, E.; Titas, M.A.; Shpilman, A.; Menzoian, J.O.; Watkins, M.T.; Raffetto, J.D.; et al. Predictive Value of Reactive Hyperemia for Cardiovascular Events in Patients With Peripheral Arterial Disease Undergoing Vascular Surgery. *Arterioscler. Thromb. Vasc. Biol.* **2007**, *27*, 2113–2119. [CrossRef]
27. Yeboah, J.; Folsom, A.R.; Burke, G.L.; Johnson, C.; Polak, J.F.; Post, W.; Lima, J.A.; Crouse, J.R.; Herrington, D.M. Predictive Value of Brachial Flow-Mediated Dilation for Incident Cardiovascular Events in a Population-Based Study. *Circulation* **2009**, *120*, 502–509. [CrossRef]
28. Kitta, Y.; Obata, J.; Nakamura, T.; Hirano, M.; Kodama, Y.; Fujioka, D.; Saito, Y.; Kawabata, K.; Sano, K.; Kobayashi, T.; et al. Persistent Impairment of Endothelial Vasomotor Function Has a Negative Impact on Outcome in Patients With Coronary Artery Disease. *J. Am. Coll. Cardiol.* **2009**, *53*, 323–330. [CrossRef]
29. Gori, T.; Parker, J.D.; Münzel, T. Flow-mediated constriction: Further insight into a new measure of vascular function. *Eur. Heart J.* **2011**, *32*, 784–787. [CrossRef]
30. Gori, T.; Muxel, S.; Damaske, A.; Radmacher, M.-C.; Fasola, F.; Schaefer, S.; Schulz, A.; Jabs, A.; Parker, J.D.; Münzel, T. Endothelial function assessment: Flow-mediated dilation and constriction provide different and complementary information on the presence of coronary artery disease. *Eur. Heart J.* **2012**, *33*, 363–371. [CrossRef]
31. Spiro, J.R.; Digby, J.E.; Ghimire, G.; Mason, M.; Mitchell, A.G.; Ilsley, C.; Donald, A.; Dalby, M.C.D.; Kharbanda, R.K. Brachial artery low-flow-mediated constriction is increased early after coronary intervention and reduces during recovery after acute coronary syndrome: Characterization of a recently described index of vascular function. *Eur. Heart J.* **2011**, *32*, 856–866. [CrossRef]
32. Cracowski, J.-L.; Minson, C.T.; Salvat-Melis, M.; Halliwill, J.R. Methodological issues in the assessment of skin microvascular endothelial function in humans. *Trends Pharmacol. Sci.* **2006**, *27*, 503–508. [CrossRef] [PubMed]
33. Roustit, M.; Cracowski, J.-L. Non-invasive Assessment of Skin Microvascular Function in Humans: An Insight Into Methods. *Microcirculation* **2012**, *19*, 47–64. [CrossRef] [PubMed]
34. Sibrecht, G.; Bugaj, O.; Filberek, P.; Nizinski, J.; Kusy, K.; Zielinski, J.; Guzik, P. Flow-Mediated Skin Fluorescence Method for Non-Invasive Measurement of the NADH at 460 nm—A Possibility To Assess the Mitochondrial Function. *Postępy Biol. Komórki.* **2017**, *44*, 333–352.
35. Nilsson, H.; Aalkjaer, C. Vasomotion: Mechanisms and physiological importance. *Mol. Interv.* **2003**, *3*, 79–89. [CrossRef]
36. Rossi, M.; Carpi, A.; Galetta, F.; Franzoni, F.; Santoro, G. The investigation of skin blood flowmotion: A new approach to study the microcirculatory impairment in vascular diseases? *Biomed. Pharmacother.* **2006**, *60*, 437–442. [CrossRef] [PubMed]
37. Aalkjær, C.; Boedtkjer, D.; Matchkov, V. Vasomotion—What is currently thought? *Acta Physiol.* **2011**, *202*, 253–269. [CrossRef]
38. Stefanovska, A.; Bracic, M.; Kvernmo, H.D. Wavelet analysis of oscillations in the peripheral blood circulation measured by laser Doppler technique. *IEEE Trans. Biomed. Eng.* **1999**, *46*, 1230–1239. [CrossRef]
39. Kvandal, P.; Landsverk, S.A.; Bernjak, A.; Stefanovska, A.; Kvernmo, H.D.; Kirkebøen, K.A. Low-frequency oscillations of the laser Doppler perfusion signal in human skin. *Microvasc. Res.* **2006**, *72*, 120–127. [CrossRef]
40. Bernjak, A.; Clarkson, P.B.M.; McClintock, P.V.E.; Stefanovska, A. Low-frequency blood flow oscillations in congestive heart failure and after beta1-blockade treatment. *Microvasc. Res.* **2008**, *76*, 224–232. [CrossRef]
41. Clough, G.F.; Kuliga, K.Z.; Chipperfield, A.J. Flow motion dynamics of microvascular blood flow and oxygenation: Evidence of adaptive changes in obesity and type 2 diabetes mellitus/insulin resistance. *Microcirculation* **2017**, *24*, 1–13. [CrossRef]
42. Semenza, G.L. Pharmacologic Targeting of Hypoxia-Inducible Factors. *Annu. Rev. Pharmacol. Toxicol.* **2019**, *59*, 379–403. [CrossRef] [PubMed]
43. Katarzyńska, J.; Zieliński, J.; Marcinek, A.; Gebicki, J. New Approach to Non-Invasive Assessment of Vascular Circulation Based on the Response to Transient Ischemia. *Vasc. Health Risk Manag.* **2022**, *18*, 113–116. [CrossRef] [PubMed]
44. Katarzynska, J.; Borkowska, A.; Czajkowski, P.; Los, A.; Szczerbinski, L.; Milewska-Kranc, A.; Marcinek, A.; Kretowski, A.; Cypryk, K.; Gebicki, J. Flow Mediated Skin Fluorescence technique reveals remarkable effect of age on microcirculation and metabolic regulation in type 1 diabetes. *Microvasc. Res.* **2019**, *124*, 19–24. [CrossRef]
45. Chudzik, M.; Cender, A.; Mordaka, R.; Zieliński, J.; Katarzyńska, J.; Marcinek, A.; Gebicki, J. Chronic Fatigue Associated with Post-COVID Syndrome versus Transient Fatigue Caused by High-Intensity Exercise: Are They Comparable in Terms of Vascular Effects? *Vasc. Health Risk Manag.* **2022**, *18*, 711–719. [CrossRef]
46. Gebicki, J.; Katarzynska, J.; Marcinek, A. Effect of psychological stress on microcirculation oscillations: Diagnostic aspects. *Vasc. Health Risk Manag.* **2023**, *19*, 79–82. [CrossRef]
47. Catrina, S.B.; Zheng, X. Hypoxia and hypoxia-inducible factors in diabetes and its complications. *Diabetologia* **2021**, *64*, 709–716. [CrossRef]
48. Zheng, X.; Narayanan, S.; Xu, C.; Eliasson Angelstig, S.; Grünler, J.; Zhao, A.; Di Toro, A.; Bernardi, L.; Mazzone, M.; Carmeliet, P.; et al. Repression of hypoxia-inducible factor-1 contributes to increased mitochondrial reactive oxygen species production in diabetes. *eLife* **2022**, *11*, e70714. [CrossRef] [PubMed]
49. Dei Cas, A.; Gnudi, L. VEGF and angiopoietins in diabetic glomerulopathy: How far for a new treatment? *Metabolism* **2012**, *61*, 1666–1673. [CrossRef] [PubMed]
50. Botusan, I.R.; Sunkari, V.G.; Savu, O.; Catrina, A.I.; Grünler, J.; Lindberg, S.; Pereira, T.; Ylä-Herttuala, S.; Poellinger, L.; Brismar, K.; et al. Stabilization of HIF-1α is critical to improve wound healing in diabetic mice. *Proc. Natl. Acad. Sci. USA* **2008**, *105*, 19426–19431. [CrossRef] [PubMed]

51. Cha, D.R.; Kim, N.H.; Yoon, J.W.; Jo, S.K.; Cho, W.Y.; Kyu, K.H.; Won, N.H. Role of vascular endothelial growth factor in diabetic nephropathy. *Kidney Int. Suppl.* **2000**, *58*, 104–112. [CrossRef]
52. Tanaka, T. Expanding roles of the hypoxia-response network in chronic kidney disease. *Clin. Exp. Nephrol.* **2016**, *20*, 835–844. [CrossRef] [PubMed]
53. Los-Stegienta, A.; Borkowska, A.; Cypryk, K. Assessment of microvascular function using a novel technique Flow Mediated Skin Fluorescence (FMSF) in patients with diabetic kidney disease: A preliminary study. *Microvasc. Res.* **2022**, *144*, 104417. [CrossRef] [PubMed]
54. Katarzynska, J.; Borkowska, A.; Los, A.; Marcinek, A.; Cypryk, K.; Gebicki, J. Flow-Mediated Skin Fluorescence (FMSF) Technique for Studying Vascular Complications in Type 2 Diabetes. *J. Diabetes Sci. Technol.* **2020**, *14*, 693–694. [CrossRef] [PubMed]
55. Gebicki, J.; Marcinek, A.; Zielinski, J. Assessment of Microcirculatory Status Based on Stimulation of Myogenic Oscillations by Transient Ischemia: From Health to Disease. *Vasc. Health Risk Manag.* **2021**, *17*, 33–36. [CrossRef]
56. Gebicki, J.; Katarzynska, J.; Cholewinski, T.; Sieron, L.; Marcinek, A. Flowmotion Monitored by Flow Mediated Skin Fluorescence (FMSF): A Tool for Characterization of Microcirculatory Status. *Front. Physiol.* **2020**, *11*, 702. [CrossRef]
57. Armstrong, D.G.; Boulton, A.J.M.; Bus, S.A. Diabetic foot ulcers and their recurrence. *N. Engl. J. Med.* **2017**, *376*, 2367–2375. [CrossRef]
58. Thangarajah, H.; Vial, I.N.; Grogan, R.H.; Yao, D.; Shi, Y.; Januszyk, M.; Galiano, R.D.; Chang, E.I.; Galvez, M.G.; Glotzbach, J.P.; et al. HIF-1α dysfunction in diabetes. *Cell Cycle* **2010**, *9*, 75–79. [CrossRef]
59. Thangarajah, H.; Yao, D.; Chang, E.I.; Shi, Y.; Jazayeri, L.; Vial, I.N.; Galiano, R.D.; Du, X.-L.; Grogan, R.; Galvez, M.G.; et al. The molecular basis for impaired hypoxia-induced VEGF expression in diabetic tissues. *Proc. Natl. Acad. Sci. USA* **2009**, *106*, 13505–13510. [CrossRef]
60. Lin, C.; Yin, G.; Ou, M.; Zheng, S. The effects of HIF-1α and VEGF on wound healing in diabetic mice. *Biomed. Res.* **2017**, *28*, 8121–8124.
61. Los-Stegienta, A.; Katarzynska, J.; Borkowska, A.; Marcinek, A.; Cypryk, K.; Gebicki, J. Differentiation of diabetic foot ulcers based on stimulation of myogenic oscillations by transient ischemia. *Vasc. Health Risk Manag.* **2021**, *17*, 145–152. [CrossRef]
62. Dawson, E.A.; Green, D.J.; Cable, N.T.; Thijssen, D.H.J. Effects of acute exercise on flow-mediated dilatation in healthy humans. *J. Appl. Physiol.* **2013**, *115*, 1589–1698. [CrossRef] [PubMed]
63. McClean, C.; Harris, R.A.; Brown, M.; Brown, J.C.; Davison, G.W. Effects of Exercise Intensity on Postexercise Endothelial Function and Oxidative Stress. *Oxid. Med. Cell. Longev.* **2015**, *2015*, 723679. [CrossRef] [PubMed]
64. Henderson, S.A.; Graham, H.K.; Mollan, R.A.B.; Riddoch, C.; Sheridan, B.; Johnston, H. Calcium homeostasis and exercise. *Int. Orthop.* **1989**, *13*, 69–73. [CrossRef] [PubMed]
65. Maïmoun, L.; Sultan, C. Effect of Physical Activity on Calcium Homeostasis and Calciotropic Hormones: A Review. *Calcif. Tissue Int.* **2009**, *85*, 277–286. [CrossRef]
66. Libby, P.; Lüscher, T. COVID-19 is; in the end; an endothelial disease. *Eur. Heart J.* **2020**, *41*, 3038–3044. [CrossRef]
67. Siddiqi, H.K.; Libby, P.; Ridker, P.M. COVID-19—A vascular disease. *Trends Cardiovasc. Med.* **2021**, *31*, 1–5. [CrossRef]
68. Østergaard, L. SARS-CoV-2 related microvascular damage and symptoms during and after COVID-19: Consequences of capillary transit-time changes; tissue hypoxia and inflammation. *Physiol. Rep.* **2021**, *9*, e14726. [CrossRef]
69. Wirth, K.J.; Scheibenbogen, C. Dyspnea in Post-COVID Syndrome Following Mild Acute COVID-19 Infections: Potential Causes and Consequences for a Therapeutic Approach. *Medicina* **2022**, *58*, 419. [CrossRef]
70. Nguyen, T.; Johnston, S.; Clarke, L.; Smith, P.; Staines, D.; Marshall-Gradisnik, S. Impaired calcium mobilization in natural killer cells from chronic fatigue syndrome/myalgic encephalomyelitis patients is associated with transient receptor potential melastatin 3 ion channels. *Clin. Exp. Immunol.* **2017**, *187*, 284–293. [CrossRef]
71. Perrin, R.; Riste, L.; Hann, M.; Walther, A.; Mukherjee, A.; Heald, A. Into the looking glass: Post-viral syndrome post COVID-19. *Med. Hypotheses.* **2020**, *144*, 110055. [CrossRef]
72. Bornstein, S.R.; Voit-Bak, K.; Donate, T.; Rodionov, R.N.; Gainetdinov, R.R.; Tselmin, S.; Kanczkowski, W.; Müller, G.M.; Achleitner, M.; Wang, J.; et al. Chronic post-COVID-19 syndrome and chronic fatigue syndrome: Is there a role for extracorporeal apheresis? *Mol. Psychiatry* **2022**, *27*, 34–37. [CrossRef] [PubMed]

MDPI AG
Grosspeteranlage 5
4052 Basel
Switzerland
Tel.: +41 61 683 77 34

Biology Editorial Office
E-mail: biology@mdpi.com
www.mdpi.com/journal/biology

www.ingramcontent.com/pod-product-compliance
Lightning Source LLC
Chambersburg PA
CBHW062100220926
51850CB00010B/2830
* 9 7 8 3 7 2 5 8 5 0 3 5 8 *